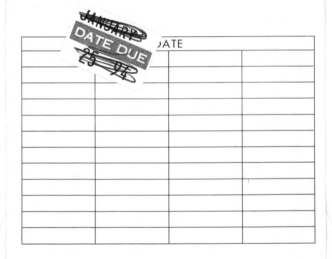

From **Machine Shop** to
Industrial Laboratory

**Johns Hopkins Studies
in the History of Technology**

Merritt Roe Smith,
Series Editor

From **Machine Shop** to **Industrial Laboratory**

*Telegraphy and the Changing Context
of American Invention, 1830–1920*

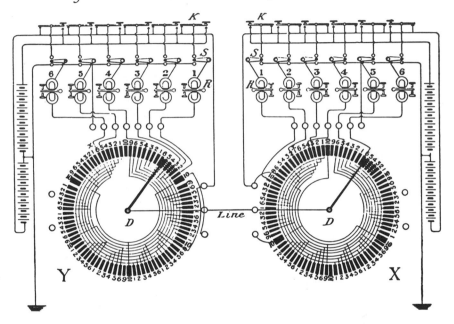

Paul Israel

The Johns Hopkins University Press

Baltimore and London

The Johns Hopkins University Press
701 West 40th Street
Baltimore, Maryland 21211-2190
The Johns Hopkins Press Ltd., London

Library of Congress Cataloging-in-Publication Data

Israel, Paul.
From machine shop to industrial laboratory: telegraphy and the changing context of
American invention, 1830–1920/Paul Israel.
    p.   cm.—(Johns Hopkins studies in the
history of technology; new ser., no. 14)
    Includes bibliographical references and
index.
    ISBN 0-8018-4379-0
    1. Technology—United States—History.
2. Telegraph—United States—History. 3.
Inventions—United States—History. I. Title.
II. Series.
T21.I77, 1992   92-812
609.73—dc20

Contents

Acknowledgments

A BOOK, like an invention, is never created alone and I would like to thank the many persons who have assisted me over the years. This book began as a paper in a seminar conducted by Prof. Reese V. Jenkins. As my dissertation director, Professor Jenkins continued to provide invaluable advice and criticism that has greatly improved the quality of my analysis. The other members of my dissertation committee, Leonard S. Reich, Richard L. McCormick, and Michael Mahoney, all offered thoughtful suggestions that have helped turn this from a dissertation into a book.

Numerous other scholars have stimulated my thinking through discussions and criticism. None have been more important than my fellow editors at the Thomas A. Edison Papers: Robert Rosenberg, Keith Nier, and Melodie Andrews, whose friendship has been as important as their other contributions. Phil Pauley offered invaluable insights and ideas that greatly improved the manuscript. Robert Friedel and Toby Appel have also been good friends as well as offering suggestions and other assistance. I would also like to thank the unknown reader for Johns Hopkins University Press whose comments proved especially useful in revising the manuscript. Finally, there are countless scholars in the Johns Hopkins Society for the History of Technology who have stimulated my thinking with their papers and through conversation.

Scholarship also requires the contribution of archivists and museum personnel. Bernard Finn at the Division of Electricity of the National Museum of American History provided assistance with the Western Union Collection and other materials, as did the staff of the museum's archives. Margaret Jasko and her staff at the Western Union Telegraph Company provided me access to records without which this book could not have been written. Crucial U.S. Patent Office records were made accessible through the assistance of John Butler and Marjorie Ciarlante of the National Archives. The staff at the Federal Archives and Records Center in Bayonne, New Jersey, helped me with the federal court records located there. Several other archival staffs assisted my efforts as well at the Smithsonian Institution Archives, the Library of Congress, and the New-York Historical Society. Peggy Weniger of the Alexander Library at Rutgers University provided invaluable help with interlibrary loan requests. Additional help with published sources was supplied by the staffs of the New York Public Library and the Library of Congress.

Two fellowships made it possible for me to take leave time from the Edison Papers in order to undertake this work: my thanks to the Institute of Electrical and Electronic Engineers for their Fellowship in Electrical History and to the University of Illinois Foundation for a Rovensky Fellowship.

A special thanks is due to my wife, Kali Israel, for her intellectual stimulation. She often heard my first tentative ideas and her questions and ideas helped to shape my analysis. Her careful copyediting also helped to improve the quality of my thoughts. Finally, to my cat, Athena, for her companionship during the long days and nights spent composing this work.

From **Machine Shop** to
Industrial Laboratory

Introduction

Every phase of the mental activity of the country is more or less represented in this great system. The fluctuations in the markets; the price of stocks; the premium on gold; the starting of railroad trains; the sailing of ships; the arrival of passengers; orders for merchandise and manufactures of every kind; bargains offered and bargains closed; sermons, lectures, and political speeches; fires, sickness, and death; weather reports; the approach of the grasshopper and the weevil; the transmission of money; the congratulations of friends—everything, from the announcement of a new planet down to an inquiry for a lost carpet-bag, has its turn in passing the wires. Amidst all this private business perhaps some political incident of prime importance transpires at Washington, or some terrible casualty shocks Boston or Chicago. Almost instantaneously the knowledge of it reaches the nearest ganglion of our great artificial nervous system, and it spreads simultaneously in every direction throughout the land.
"The Telegraph"
Harper's New Monthly Magazine (1873)

I T HAS BECOME commonplace to refer to ours as the age of communication and information but, as the above passage suggests, nineteenth-century Americans experienced their own revolution in communications and information with the advent of the telegraph. In his book *Media and the American Mind from Morse to McLuhan*, communications historian Daniel Czitrom noted that public reaction to the first electric communication system evidenced a recognition of profound changes in human communications. In beginning his study with the telegraph, Czitrom acknowledged the important historical context it provides for analyzing modern communications technology. Although most scholars of communication technology have paid little attention to telegraphy, they do agree with Czitrom that it represented an important transition from print to electric communication media.[1]

What has been less clear to scholars are the important ways in which telegraphy also stood at the center of another major transition in American society—the emergence of corporate capitalism and its reorganization of technology research and development. While historian Alfred Chandler has described the railroads as the first modern corporations, it was Western

Union Telegraph Company that emerged as the first corporation to operate on a nationwide scale and to draw attention as a dangerous new form of monopoly. Western Union also supported Thomas Edison's Menlo Park laboratory, which most historians agree "pointed the way toward the systematic research of the technological age" and served as a transitional type that helped transform American invention.[2] Edison, who began his career in the telegraph industry, is himself recognized as a transitional figure standing between the independent inventor and the modern industrial researcher.

Recent historical research has extensively broadened our understanding of the rise of modern corporate-based industrial research laboratories in the twentieth century, especially among those companies—primarily electrical and chemical—that have established science-based research traditions.[3] This literature has been less successful in explaining nineteenth-century patterns of invention or in exploring alternative forms of twentieth-century industrial research. At the same time, most works on early American technology discuss invention primarily in relationship to larger issues of technical change in particular industries. Nonetheless, it is possible to find in this diverse work some general outlines for a new history of American invention. The connecting thread in these studies is the role of the workplace in invention—more specifically, the machine shop as a central institution for the development of new technology. In the place of many lone, independent inventors we begin to see a pattern of cooperative shop invention in which skilled operatives, superintendents, machinists, and manufacturers make up technological communities that draw on practical experience to design, build, and refine new technology.[4]

One of the few studies specifically addressing invention in early America, Brooke Hindle's *Emulation and Invention,* suggests that this style of invention was made possible by the mechanical character of so much nineteenth-century technology. Machinery consisted primarily of gears, ratchets, cams, escapements, and other mechanisms whose construction and mechanical movements could be readily examined on the shop floor. As a result, machine shops became the schools and machines the books of an American technology that many at the time considered the embodiment of democratic values. The self-taught mechanic was Everyman, capable of improving the technology with which he worked. An expanding manufacturing economy and growing technological enthusiasm embodied in the nation's patent system provided further stimulus for invention. Hindle's work also hints at the importance of this mechanical tradition in the early telegraph industry; telegraph inventor Samuel Morse was a major subject of Hindle's book.

Telegraphy evolved its own electromechanical shop tradition influenced by the more general shop tradition in American invention. One of the reasons for this is the nature of telegraph technology itself. Although telegraph instruments were powered by electricity they incorporated mechanical elements that reinforced the importance of practical experience and machine-shop design practice. Because shop invention was so closely tied to the workplace, any attempt to understand its influence in telegraphy also involves an analysis of the ways in which workers acquired skill and knowledge, as well as an examination of how and why they applied it to invention. Both the telegraph operating room and the telegraph manufacturing shop provided informal technical schools and experimental laboratories similar to those found in other industries.

Telegraphy was influenced by new scientific knowledge of electricity as well. Indeed, without prior scientific research, the use of electricity for communications was unthinkable. Further, it is clear that by the end of the century the source of knowledge about electricity was shifting from practical experience and self-education to formal, scientifically grounded engineering programs in colleges and universities. Historians interested in the rise of industrial research laboratories in the twentieth century have demonstrated the crucial role played by new scientific knowledge in their creation.[5] Although the telegraph industry and its electromechanical shop tradition initially influenced the new electrical industries of telephony and electric lighting, the greater centrality of science in the newer fields caused them to soon diverge in their pattern of technical development.[6] Important differences in the scientific knowledge base required for telegraphy helped to keep the shop tradition viable as it became outmoded in other fields. This raises important questions regarding our understanding of the modern history of industrial research in other industries that remained grounded in the mechanical technology that gave rise to this shop tradition.

The ultimate decline of shop invention in telegraphy is tied to changes in the workplace which began to alter the sources of technical knowledge. As occurred in other industries, the machine shop, which had served as the laboratory of shop invention, was transformed by the new techniques of scientific management. Developed by engineering professionals whose belief in scientific rationalization merged with their own status insecurities, scientific management led to a transfer of knowledge from the worker on the shop floor to the engineers and managers of new corporate bureaucracies. The growing influence of engineering in telegraphy produced its own changes in the operating room. While not consciously designed to redefine operators' skills, they nonetheless had that effect. The larger change in workplace relations that occurred at the turn of the century thus becomes an important subject of study when examining the changing locus of inventive activity.

These changes in the workplace were themselves part of a larger phenomenon growing out of the increasing scale and complexity of American economic activity—the rise of the modern, large-scale, bureaucratically organized corporation. It was in the telegraph industry that one of the first modern corporations appeared, as well as the first national company to achieve a near monopoly. The growth of monopoly capital in the form of the corporation provided the setting not only for scientific management but also for the industrial research laboratory. Both phenomena are related to the growth of a managerial imperative that altered the structure of American industry and invention.

The telegraph industry's transitional character makes it an appropriate vehicle for exploring the shift from shop invention to industrial research. Although telegraphy required new scientific information for its birth, its practical introduction and future development grew out of the mechanical shop tradition of invention. At the same time, the industry was integral to the growing scale and complexity of the American economy and helped create a new corporate context for American industry that would reorganize inventive activity. Indeed, the telegraph industry was responsible for one of the first institutions that helped to shape this reorganization by transforming shop invention. This, of course, was Edison's Menlo Park laboratory, which physically divorced invention from manufacturing, although it retained the skilled workmen and tools of the machine shop. Like that laboratory, the telegraph industry only partially altered the pattern of American inventive activity, but together they created a new pattern of invention that would be more fully developed in other industries and other institutions.

This new pattern of invention has evolved its own scientific traditions in place of the older shop culture and become synonymous with modern industrial research.[7] But the history of shop invention remains of more than antiquarian interest. Recently, scholars, industrial leaders, and others concerned with issues of competitiveness in a global economy have begun to examine again the role of the workplace and of shop knowledge, particularly in regard to "the utilization and diffusion of the results of scientific and engineering research."[8] While recognizing that science-based research has had a profound impact on modern technology, they have begun to pay new attention to the relationship between technological change, manufacturing, and design. It is this relationship, central to nineteenth-century inventive practice, that is examined in this book.

1

The Rise of American
Mechanical Invention

INVENTION emerged as a significant factor in American life in the years between Samuel Morse's birth in 1791 and his early inventive work on the first American telegraph in the 1830s. Those years marked the formative period of American ideas about invention as well as the organization of inventive activity within the society and its integration into the nation's economic system. During this period cultural beliefs regarding the nature of invention and its relationship to social progress became embodied in the newly created American patent system. The patent system also incorporated ideas concerning the connection between new technology and the nation's economic development. The course of American economic development provided its own framework by helping to organize the institutional setting and technical patterns of inventive activity. The structure of American invention as it developed from 1790 to 1840 had important consequences for Morse and his contemporaries as well as for the inventors who followed them. Before turning to telegraphy it is essential to provide an overview of these formative years and examine the factors that influenced the development of nineteenth-century American mechanical invention.

A Social Environment for Invention

In the early republic, burgeoning economic growth, an expanding manufacturing sector, and abundant natural resources helped to establish an American belief that the nation's social progress and democratic institutions were best ensured by continued material progress. Because technology enabled manufacturers to produce more and facilitated the exploitation of the country's abundant resources, Americans saw it as an important source of material progress. They therefore invested technology with a special role "in providing the physical means of achieving democratic objectives of political, social, and economic equality."[1] At the time of Morse's birth, Americans were ambivalent about the benefits offered by material progress achieved through technological change. But as changing political and economic conditions transformed the agrarian society of Morse's youth into an emergent industrial power, they came to embrace new technology as a positive social goal. In the process, Americans came to celebrate and promote invention.

When Samuel Morse was born in 1791 the young nation was imbued with republican values molded by the leaders of the War for Independence. Based on their ideas about classical antiquity, Americans generally regarded republics as inherently fragile. The stability of this form of government rested on the virtuous character of an independent citizenry, but this virtue could be corrupted as material prosperity increased luxury and wealth among some members of the society and fostered the growth of special interests susceptible to government power and tyrannical rule.[2]

To sustain their own republic, Americans sought to organize their society in ways that would produce an independent and virtuous citizenry. Most agreed with Thomas Jefferson that yeoman farmers, by virtue of their land ownership and their self-sufficient household production of goods, were the ideal republican citizens. Many argued that independent artisans, who produced only necessary goods that were not susceptible to the whims of fashion, would also maintain their autonomy and virtue. Only a minority, however, envisioned a place for large-scale manufactures in the republic. For most Americans, the example of British manufacturers, who relied on government subsidy and support, as well as a dependent lower class, demonstrated the susceptibility of industry to corruption.[3]

While generally rejecting manufacturing on the British model, Americans accepted much of the new technology that made it possible. They did so by casting this technology in a conservative light, believing that it could help preserve the conditions in American society that made possible a citizenry of independent farmers. Thomas Jefferson, who introduced new agricultural improvements at Monticello and advocated their adoption by American farmers, argued that such improvements could increase production and assure self-sufficiency. He encouraged new manufacturing technology for much the same reason. The new inventions that were enlarging the scale and significance of manufactures in Europe would be adapted to decentralized, traditional household manufacturing in America and become the means by which domestic production of goods was increased without the necessity of introducing factories. Jefferson set an example at Monticello, where he introduced textile and nail-making machinery. George Washington joined Jefferson as an ardent supporter of new inventions and also adopted improvements in agricultural and household production at Mount Vernon. Both, however, failed to recognize the essential role that slavery played in their efforts to achieve self-sufficiency in household production.[4]

Americans who advocated the adoption of new technology argued that the future prosperity and progress of their country relied on widespread knowledge of the practical arts and sciences, just as their country's political advancement emanated from knowledge of the political sciences. This faith in technological progress was a product of the same forces that promoted British industrialization during the period of the American Revolution. The Newtonian scientific revolution of the seventeenth century helped to reinforce the emergent capitalist, free market economy that emerged from the longterm breakdown of Britain's feudal economic system. The new economic system supported industrial development, and the mechanical philosophy of Newtonian science imbued British entrepreneurs acting in this environment with a belief that their knowledge of natural forces could be applied through mechanisms to provide a powerful tool for economic development.[5]

Newtonian science was disseminated throughout British elite culture by itinerant lecturers and in meetings of philosophical and scientific societies. Scientific culture spread among elites not only in provincial cities such as Manchester and Birmingham but also in Boston and Philadelphia. Although on the periphery of this culture, the leaders of American society were nonetheless influenced by scientific learning and became inculcated in the practice of mechanical thinking. In America this learning was initially confined largely to the universities, thus influencing noted religious thinkers such as Increase and Cotton Mather, who learned about the new science at Harvard and incorporated versions of it into their religious outlook. The scientific connections between British and American elites was direct in the case of Thomas Jefferson, who learned natural philosophy and mathematics at the College of William and Mary from William Small. Small later returned to Birmingham where he joined with Matthew Boulton, James Watt, and Erasmus Darwin in that city's Lunar Society. Benjamin Franklin, though largely self-taught, gained renown for his scientific work and became a member of the Royal Society of London. Local elites in both Britain and America engaged in programs of self-improvement in which science was a central component. They attended lectures, read and published books and journals, and formed scientific and philosophical societies. The societies established in provincial cities, though they produced few important discoveries in "pure" science, were essential for promoting improvements in agriculture, manufacturing, and the useful arts.[6]

Members of these scientific societies were imbued with the empirical and utilitarian ideals of Francis Bacon who had advocated the use of scientific knowledge for solving practical problems. Their highly practical vision of science led them to promote industrial and agricultural improvement in both countries. In America, scientific societies promoted such improvements in a wide variety of ways and sometimes directly supported the work of inventors by offering them premiums or other awards and by publicizing their endeavors. Promotional efforts by individuals also stimulated interest in new inventions. Prominent Americans such as Washington and Jefferson encouraged and sometimes actively supported particular inventions. All of these efforts helped create a more positive environment for technological change in the new nation.[7]

This positive environment was also made possible by an agricultural community whose familiarity with machinery and tools went beyond those used directly in farming. It was the rare village that did not have a gristmill; many mills preceded town settlements. The miller was a valued member of the community and various inducements, including prime farm land, were often extended to encourage him to settle and remain in a village. These mills "often became the entering wedge of a slowly emerging market economy."[8] In some areas gristmills were also converted to sawmills during parts of the year and, particularly in New England where lumbering was an important industry, many sawmills were established in their own right. Fulling mills, used in the production of cloth, also made their appearance. The village blacksmith was a vital member of most communities, the one who made the numerous pieces of hardware required on the farm and in the mills. Some farmers also had shops in which they did much of the repair work on their farm buildings and implements and acquired a variety of skills. Those farmers who exhibited particular skill as craftsmen might undertake projects for their neighbors and in the process build up a flourishing trade, eventually converting farming to a part-time activity. In other instances skilled craftsmen from Europe who became farmers in the New World continued to practice their former trades on a part-time basis. Much of the skilled and semi-skilled workforce of the early nineteenth century emerged from these rural communities. Although farmers and rural craftsmen were technologically conservative during the colonial period, by the end of the eighteenth century increasing European demand for American foodstuffs encouraged commercial farming, creating a new emphasis on agricultural specialization. This in turn fostered a growing interest in technological and scientific improvements to increase both farm and manufacturing productivity.[9]

Americans especially sought improvements that increased commerce and opened up western lands to agricultural production for the country's growing population. While new farmland would enable America to retain its agrarian character, the accompanying benefits of increased commerce were less certain. Although classical republicanism viewed foreign commerce as a potentially corrupting influence, Americans who were intimately engaged in the international economy transformed it into a virtue. Proponents of foreign trade argued that it could ensure that Americans remained industrious, and thus virtuous, by providing markets for the increasing agricultural goods produced as new lands were opened to cultivation. They also argued that foreign commerce would fortify the agricultural character of the republic in other ways. While hoping that household manufactures would meet domestic demand, some American leaders recognized that these were insufficient and therefore hoped that foreign trade could obviate the need to establish potentially corrupting home manufactures. The desire to see virtue in foreign commerce was reinforced by rising demand for American agricultural goods created by the European wars following the French Revolution. By ending mercantile restrictions to allow free trade, Americans hoped they could foster interdependence among nations, thereby improving international relations and reducing the potential for war.[10]

America's involvement in the European wars not only dashed those hopes, but also promoted the integration of manufacturing interests into the republican vision as an essential ingredient in the maintenance of national liberty. This occurred as Americans engaged first in embargo and then in war between 1807 and 1815, and began to recognize the dangers of relying on foreign trade for necessary goods. As a result, some began to urge the development of a mixed economy of agriculture and manufacturing in order to sustain the new nation's economic independence. As manufacturing became perceived as essential to the preservation of the American republic, it was argued that those engaged in such enterprises would depend on the maintenance of the republic for their livelihood and therefore acquire a stake in society equivalent to that of the yeoman farmer's land.[11]

Manufacturing was also supported by the liberal ideals being integrated into the republican outlook by Americans committed to an expansive commercial economy. They began to argue that the free pursuit of individual economic interest was the underpinning of a political liberty in which private gain could ensure the public good. Seeing the nation's social structure as ordered by economic rather than political relationships, commercially minded Americans asserted that economic growth and material prosperity would foster social equality by making more goods and greater material comfort more widely available within the society. Furthermore, they contended that the consumption of manufactured goods, including luxuries, was a crucial incentive for a society of industrious producers. Some advocates now lauded increased luxury and comfort as important incentives to invention and economic growth. Economic growth in turn became the basis for a new progressive view of history that began to supplant the classical republican belief in inevitable decay. Proponents of economic growth argued that industrious and productive citizens spurred by the desire for private gain would create new wealth from the abundant resources of the American continent, thus ensuring the nation's future prosperity. Technology played a crucial role in providing the means to exploit the nation's resources.[12]

This liberal reinterpretation of republican society and the place of manufactures in it encouraged a redefinition of the virtuous citizen as an industrious producer, thus making room in the new republic for those engaged in manufacturing, including factory workers. Advocates of manufacturing and technological progress argued that American ingenuity would produce labor-saving machinery to make possible factories requiring only a temporary workforce drawn from the marginal members of society—young women, children, and the aged. Their temporary employment in such enterprises would instill in these workers the sense of industriousness that was now seen as a source of republican virtue. By means of technology, America would escape the development of a dependent working class as was occurring in Europe.[13] In the opening decades of the nineteenth century debates over the emergence of factories and the nature of their workforce thus helped to wed technological innovation to an emerging republican ideology of manufacturing as well as to the maintenance of national independence and liberty. Such technological visions also became merged with entrepreneurial strategies and general American beliefs in scientific improvement to encourage invention throughout the society.

Incentives for Invention

Although the young republic created a general environment conducive to invention, inventors found their efforts frustrated by government policies and by the slow growth of a manufacturing market to produce and use their inventions. During the eighteenth century, British patents granting limited monopolies for new inventions became common, though they were little used in the American colonies. After the War for Independence, the federal government of the Confederation period offered no patent protection, relying on those granted by the individual states. Because states granted patent protection for different periods of time, American inventors found it necessary to acquire several patents to protect their inventions. As a result, they were also faced with the burden of pursuing patent infringers within each state. To secure more certain support for their efforts, some inventors attempted to acquire direct subsidies from state or federal government. Government awards, however, were a delicate political issue and were granted in an arbitrary fashion. All of these problems came together as claimants to the invention of the steamboat disputed priority and sought public support.[14]

The issues raised by steamboat inventors John Fitch and James Rumsey helped to encourage the Constitutional Convention to adopt a clause supporting the establishment of a federal patent system. The reorganization of the federal government under the Constitution, which was designed to promote more harmonious and regular economic and political arrangements between the states and the federal government, gave the federal government power "to promote the progress of science and the useful arts by securing for limited times to authors and inventors the exclusive right to their respective writings and discoveries."[15] The first Congress assembled under the new Constitution promptly passed the new nation's initial patent law in 1790.

As head of the first Board of Examiners, Secretary of State Thomas Jefferson subjected patent applications to a rigorous examination in order to insure that only new and useful inventions were granted patents.[16] His general dislike of monopoly, however, made him extremely cautious in allowing patents, and the slow process of examination, exacerbated by the limited time available to the busy cabinet officers who made up the board, caused dismay among many inventors. Jefferson felt that the patent law should primarily be concerned with the wide dissemination of knowledge and only secondly with assuring profit for inventors. Jefferson opposed the granting of the patent monopoly to any application that claimed underlying principles for a class of inventions, believing that such principles were the product of accumulated knowledge contributed by many persons. He also opposed patents that used known devices in new ways, a common feature of technological innovation.[17] Jefferson recognized that the limited monopoly provided in the patent law encouraged inventors, but his actions as head of the patent board reflected classical republican ideals in supporting an enlightened citizenry and opposing monopolies.

Intimately involved in the marketplace and imbued with emerging liberal ideas, inventors complained that the examination system was too slow and did not meet their needs. Inventors argued that their work was in the public interest, but rejected Jefferson's view that there should not be property in ideas. They therefore insisted that patent protection be made more widely available so that they could trade their inventions freely in the market, which would determine the value of their ideas. In 1793, they achieved a more open system of patent protection, which required only that an invention be registered. Because Congress was concerned that it would require the creation of a federal bureaucracy, it decided against improving the examination system and instead left the determination of a patent's validity to the courts. As a result, the new system provided little secure protection for an invention. Nonetheless, by 1802, a growing number of patent applications caused the creation of a separate patent office.[18]

William Thornton, patent commissioner from 1802 to 1828, was an enthusiastic champion and promoter of American invention and of the patent system. Through mechanics' journals and other publications he spread knowledge of how the system worked. He also sought means to make patent models accessible in order to encourage emulation and to increase the assistance that the office could provide to inventors. However, through the 1820s Congress was loath to provide funds for such purposes. Under Thornton's vigorous direction the system worked well enough that most Americans considered it the major reason for the increasing number of inventions being produced by their compatriots. After Thornton's death in 1828 the administrative and financial inadequacies of the patent office became more apparent under the pressure of an increasing number of patent applications. Furthermore, even though Thornton had sought means to prevent the issuance of fraudulent or worthless patents, Congress did not respond. As a result, inventors were faced with expensive lawsuits to protect their rights and the public was beset by patent speculators. Consequently, the early 1830s saw a deterioration of confidence in the patent system and increasing agitation for its reform, particularly as the industrializing American economy increased the value of patents.[19]

The reform of the Patent Office was part of a general reform of the federal government conducted by the Jacksonians, who responded to and fostered a general democratization taking place within American society. They sought to overturn an older republican style of government by elites, in which public-spirited men of great talent such as Thornton dedicated their lives to government service. In its place they sought to erect a more egalitarian system in which individuals of common intelligence and talents served short terms in offices operated under a set of codified regulations. Although the growing importance of public patronage in Jacksonian political parties also helped to prompt this reform and later gave rise to scandals that touched many government offices, including the Patent Office, the reformers believed they were returning the conduct of government to the constitutional principles of the rule of law.[20]

The major reform of the patent system involved the reinstitution of patent examinations. Now, however, the process was conducted by lower-level officials rather than by cabinet officers, with sufficient staffing to process the applications. Examiners inspected patent applications for novelty, originality, and utility, thus providing greater insurance of the validity of issued patents and reducing the possibilities of fraud. Although recognizing that invention was a cumulative process, Americans nonetheless regarded it as an intermittent and discontinuous activity undertaken by inspired individuals. The examination system sought to protect the rights of such individuals and was expected to mediate between the inventors' concerns that intellectual property be protected and the requirement that new ideas be publicly disclosed.

The 1836 patent law helped reinforce the intimate link that had been formed between technological and material progress and democratic goals. Because the law required public access to an invention in exchange for the incentive of a temporary monopoly and the possibility of individual financial reward, the community was expected to gain a benefit it might otherwise lose. In one of the first treatises written on the new patent system, attorney Willard Phillips expressed some reservations about the examination system. He nonetheless recognized that it better secured the rights of inventors, thus strengthening this system of exchange and furthering the advancement of the useful arts. In words that became increasingly common, Phillips concluded that by diffusing useful knowledge and encouraging invention, the American patent system would bring a greater "amount of necessaries, conveniences, comforts, luxuries and amusements, within reach of every one, for the same expense."[21]

While the patent system was perceived as the most important incentive for invention, the growth of manufacturing and the introduction of new production methods provided the greatest inducement for inventors. Those who relied on traditional craft production found themselves in an increasingly competitive environment, as manufacturing gained wider acceptance within society and the number of manufacturing concerns grew. The values of the marketplace began to overturn the centuries-old craft system. Master craftsmen and other manufacturing entrepreneurs began seeking new markets, expanding their operations, experimenting with new forms of credit, employing cheaper labor and new machinery, and engaging in inventive activity as they strove to make their businesses more efficient and profitable.[22] It was in these innovative manufacturing shops that a distinctive American style of invention began to take form and that many of the early mechanical inventions of the nineteenth century were developed.

As mobility within the ranks of the crafts became separated from the traditional apprenticeship system, masters and some journeymen began to invoke the values of self-improvement that by the 1820s pervaded American society. Urging self-discipline and education, advocates of self-improvement argued that virtue, identified now with industriousness, and knowledge were the keys to success. Useful knowledge, primarily defined as scientific and technical, was considered essential for those engaged in manufacturing. Through the acquisition and application of such knowledge individuals would progress in their crafts. As they became master craftsmen they would also help to improve the crafts themselves.[23]

As Brooke Hindle has demonstrated, the traditional mode of learning in the mechanical arts—emulation—was transformed into a dynamic source of inspiration for inventors. "Just as the best work was held up to the apprentice as a model, so invention might be encouraged by holding up as models the best inventors."[24] Already Americans had created a pantheon of inventors, including Benjamin Franklin, Robert Fulton, and Eli Whitney, which provided models for others. And the patent system itself, by making available actual physical models of patented inventions, inspired inventors to produce designs of equal or superior value to the nation. As improvements in the crafts became cumulative, they would help American society achieve greater prosperity and social progress. At the same time, invention was lauded as a means of social mobility for the mechanic.

Though conceived in egalitarian terms, for a large percentage of the population inventive activity was neither encouraged nor expected. While it was anticipated that ambitious male workers could use their mechanical skill to become superintendents and even independent entrepreneurs, women were neither expected to acquire such skill nor to turn it to invention. The very definition of invention in nineteenth-century America focused on machine design and, as Autumn Stanley has shown, the bias that existed toward mechanical invention was deeply rooted in a view of invention as a male prerogative. In her study of the list that U.S. Patent Office clerks compiled for the agency's centennial in 1890, Stanley discovered that not only did they fail to identify a large number of women inventors, but that "*machines turned out to be the single largest category of the omitted inventions*" (emphasis in original). This was especially true of machines not related to women's traditional domestic roles.[25] Women were seen as mere "machine tenders"; their experience with machinery was thought to be temporary, ending when they married and entered into the domestic sphere. While occasional notice was paid to the proposition that women were capable of invention, most commentators assumed they were speaking to a male audience when they urged mechanics to contribute to society's material progress as well as their own social mobility through invention.[26]

The primary vehicles designed to aid the mechanic in his quest for self-improvement were the mechanics' institutes established in a number of American cities. The breakdown of traditional craft organization placed the responsibility for education increasingly on the individual and these institutes served an important educational function during the 1820s and 1830s, before the era of common school reform. Indeed, the educational function of the institutes helped to further the breakdown of traditional craft practice by reducing the authority of masters in the transmission of knowledge. Although the largest and best known institutes were established by master mechanics, journeymen established their own organizations. Both agreed on the importance of self-improvement and useful knowledge in advancing the interests of their members. Toward this end the various institutes published magazines, provided libraries, and conducted lectures that disseminated information on scientific principles and technical improvements. They also encouraged craftsmen to adopt and improve new technology and to perceive invention as an important means for improving one's business and increasing social mobility.[27]

Not all craft workers or masters adopted the ideology of entrepreneurship promoted by the mechanics' institutes, but it did become increasingly dominant.[28] Stephen Vail, a Morristown, New Jersey, blacksmith, was typical of entrepreneurial craftsmen who became early manufacturing pioneers. He had been apprenticed to a village blacksmith in 1796 at the age of 16 and was later employed in a nail-cutting factory. By the age of 20 he had his own flourishing blacksmith shop, and in 1808 he joined with three partners to purchase a slitting mill situated on the site of a pre-Revolutionary sawmill. Under Vail's direction the original slitting mill, where iron bars were rolled flat and cut into strips for such items as nails, barrel hoops, and iron tires for wagons, was enlarged and improved. Although he faced initial financial difficulties, by 1815 Vail was able to buy out his partners. Vail showed an enterprising spirit in running the ironworks. In 1818, he agreed to build a steam engine for a new steamship company established by Capt. Moses Rogers to ply the Atlantic. Although his ironworks had no experience in building engines, Vail acquired the services of steam-engine designer Samuel Carson, who had been trained by the sons of steam engine innovators James Watt and Matthew Boulton. The engine was built and successfully installed on the *Savannah,* which began the first steamship voyage to Europe on 22 May 1819. Although the journey did not mark the beginning of regular steamship passage across the Atlantic and Vail never collected the money due for the engine, the engine-building project enhanced the reputation of his ironworks, which subsequently did an extensive nautical business.[29]

Some individual entrepreneurs such as Vail manufactured the products of inventors, but many undertook inventive work themselves in an effort to improve their position in the market. Marjorie Ciarlante's analysis of eminent American inventors from 1790 to 1849 indicates that among those inventors who achieved notable success, manufacturers constituted over 50 percent of the total and that those who were either self-employed or owners of a firm produced the largest percentage of patented inventions. The majority of the inventive elite focused their creative work on those inventions most closely associated with manufacturing and manufactured products. Located primarily in New England and the Middle Atlantic states, over 50 percent of them worked in cities or towns with populations over 5,000 at a time when 85 percent of the population lived in rural areas. These inventor-entrepreneurs were on the cutting edge of industrialization and often focused their activity on key inventions, such as steam engines or machine tools, related to the new production technology. Their patenting activities also reflected the shift from wood to metal products taking place in American industry as increased mechanization required more precise metal machines and machine tools.[30] The expanding metalworking industries also served as the center of American inventive activity.

A Setting for Invention

As entrepreneurs in the traditional crafts and in the new manufacturing industries produced changes in the workplace, they also created a working environment conducive to invention.[31] Much remains to be learned about the impact of industrialization and mechanization on individual craft workers, but recent studies indicate that the new production techniques created a variety of labor skills and a diversity of work environments. Before 1850, most industrial production actually took place in manufactories, small mechanized shops utilizing a variety of machines and generally employing fewer than twenty people, mostly skilled workers, many of whom aspired to ownership. The factory environment of the textile industry, with its highly organized work discipline and preponderance of semi-skilled machine operatives, was less common in other industries where skilled workmen used hand and machine tools to forge their products. They were assisted by apprentices and unskilled laborers, many of whom eventually became skilled themselves through an informal apprenticeship system.[32]

A number of those entering the manufacturing workforce came from the nation's farms. Many machinists first gained experience with mechanisms on farms and in small village shops. Gristmills, sawmills, drills, cultivators, seeders, clocks, and other machines were prevalent throughout the society and most of these devices were made of wood rather than metal. Furthermore, woodworking machinery often "demanded a higher grade of ingenuity and skill in their construction than machines for cutting and shaping metal."[33] Unlike the craft practitioner, whose skills were highly specialized, the farmer or rural craftsman had to be something of a jack-of-all-trades. The mechanical skills learned in rural America were readily transferred to the machine shop.[34]

The emerging skilled labor force was most evident in the new metalworking trades that produced machinery for American industry. At the center of the new trades were machine shops and their skilled operators. The machinist was usually proficient in the operation of all the machines in the shop and capable of making, repairing, and often designing a variety of machinery. The increased use of machinery in manufacturing placed great value on the skills of machinists, who experienced significant geographic and occupational mobility. Machinists who moved in order to improve their situation transmitted new techniques and tools between regions and industries. Some machinists eventually opened small shops of their own, a few of which grew to prominence.[35]

Early machine shops often appeared as appendages to factories for the building and repairing of the machinery they used; many such shops were connected to textile mills. Some shops also undertook contract work from the outside. Independent shops were established to build the steam engines used both for steamboats and railroads, and to power factories. Others emerged in conjunction with the arms industry, as government support for interchangeable-parts design encouraged the use of precision machine tools. With the introduction of new machinery into other industries, machine shops became common in cities and towns across the country, usually undertaking special orders and acting as "an experimental laboratory which developed and perfected industrial and mechanical processes and equipment."[36]

Stephen Vail's Speedwell Iron Works, destined to later play an important role in the early invention of the telegraph, was representative of these early American manufacturing enterprises. Most such establishments arose in communities known to historical geographers as "central places," which provided commercial, transportation, processing, or public administration facilities for the surrounding areas. Particularly in those towns with good transportation and processing facilities, mills, blacksmiths shops, and forges provided both the skill and the capital necessary for manufacturing endeavors.[37] Vail's ironworks was located on the outskirts of Morristown, New Jersey, the county seat of Morris County, which was a processing center for farm products and for the county's rich deposits of iron ore. Located on the Whippany River, the mill was connected by an extensive road network north to the iron mines and east to New York City, some thirty miles away. Under Vail's direction, Speedwell became a major producer of iron articles for mills, boats, pumps, engines, and other machinery for the New York market. Like many other metalworking firms, the shops of the Speedwell Iron Works also became an informal school disseminating technological knowledge.[38]

Technological Knowledge and Invention

As Brooke Hindle has explained, technological knowledge at this time was largely tactile and visual in nature.[39] As a result, such knowledge was primarily communicated through the direct experience of working with and observing both machines and skilled craftsmen. Indeed, the early development of American industry was retarded by the lack of skilled craftsman as well as by antimanufacturing ideologies. When new industries such as textiles and chemicals did emerge, they relied on immigrant craftsmen and mechanics, and on machinery, plans, or drawings smuggled from abroad. The canal and steam engine developments that revolutionized transportation and created internal markets for new industries likewise depended on individuals with the requisite knowledge and skills to carry out such projects and to train American craftsmen in the new techniques. Important new techniques in older industries such as mining and ironworking were also imported by immigrants.[40]

The customary manner in which knowledge was transmitted in the crafts provides another example of such transfers. In the process of emulation described by Hindle, apprentices learned by copying the work of masters, while journeymen sought to go beyond such models to create their own masterworks. These forms of emulation were an important form of technical education and became a stimulus to invention in the nineteenth century. Eugene Ferguson recounts similar education processes in his discussion of the rise of American mechanical "know-how." Sociologist Douglas Harper, studying work in a contemporary small shop, not only describes how technical knowledge was transferred by such means, but also suggests that technical skill is invested in the body and tactile senses of the craftsman. Harper calls this "working knowledge" in his book of the same title.[41]

Although traditional craft practice consisted of specialized knowledge invested in individuals who carefully guarded their techniques, the rise of machinery and mechanical technology fostered a new openness in the transmission of techniques. This openness was facilitated by what Nathan Rosenberg has called "technological convergence." The wide use of metalworking machinery meant that similar processes were common to a broad range of industries, and that machinists and machine tool companies were able to apply their skills and techniques to problems affecting many industries. In this way technological knowledge became more openly and widely diffused.[42] Because machine tools and techniques were adaptable to a number of industries, design became an interactive process in which machinists collaborated with those who used the machines in their factories and workshops.

Studies of several American industries of the first half of the nineteenth century suggest the centrality of machine shops and skilled machinists in the propagation of inventions and innovations. The growth and diversity of American manufacturing industries produced a wide field for mechanical ingenuity, although the role played by the machine shop differed from industry to industry. For example, the initial development of shoemaking machinery was undertaken by sewing machine manufacturers seeking to diversify in the 1850s, while subsequent mechanization was largely the product of a dynamic formed by competition among shoe machinery manufacturers, whose shops played a central role in inventive activity. Indeed, the location of inventive activity in the industry was directly influenced by the concentration of shoe machinery manufacturers in Massachusetts. On the other hand, major inventions in the Berkshire papermaking industry tended to develop in the paper mills, while the nearby machine shops constructed the new machinery and made small improvements. In other places, notably Philadelphia, new papermaking machinery often emerged from the machine shops. In the textile industry, invention occurred in both mills and machine shops, but in most cases the new machinery was developed in the shops. Improved steam-engine and firearm designs were largely produced in the shops of manufacturers.[43]

Even when they did not generate the initial concept for an invention, machine shops played a key role in transforming inventors' ideas into actual devices. The machinists who worked on these devices often improved inventors' designs. For example, Steven Lubar has demonstrated how John Howe's design for his rotary pinmaking machine was largely influenced by the machinists in the shop of Hoe and Company, which produced not only Robert Hoe's rotary printing press and his circular saw, but also did contract work on a host of other inventions.[44]

Skill and design often became embodied in the machines themselves, allowing one to learn design by studying machines.[45] New designs emerged through the manipulation of mechanisms such as gears, levers, cams, ratchets, and drive shafts, which provided the common vocabulary of machine construction and operation. The craft tradition of emulation, combined with a recognition that machinery incorporated tactile and kinetic features, led mechanics' institutes to promote the display and study of models and working machinery and the U.S. Patent Office to require that models accompany patent applications. In this way it was hoped that this common vocabulary would be made readily accessible.[46]

By requiring drawings as well as models, the Patent Office also acknowledged that visual representation was crucial for understanding how a mechanical device or circuit worked. The 1836 patent law specifically required that copies of patent specifications and drawings be made available to any applicant who paid the necessary fees for copying. The importance that drawings and plans played in the transmission of technological information prompted mechanics' journals and institutes to encourage the acquisition of drawing and drafting skills.[47] Most inventors had some facility for both reading and making drawings, which mechanics used in constructing working instruments and machines.

Visual images were central to the discussion of mechanical technology and a growing technical literature provided a valuable source for visual study. Particularly by mid-century, printing technology enabled journals and books to include extensive reproductions of machinery and mechanical movements. New printing technology also enabled the Patent Office to make its drawings more readily accessible. Early in the century journals like the *Useful Cabinet,* the *Boston Mechanic,* and the *Journal of the Franklin Institute,* and books such as Oliver Evans's *Young Mill-Wright and Miller's Guide* and Jacob Bigelow's *The Useful Arts,* included cuts reproducing mechanical elements. By the 1830s, tables of mechanical movements provided important information that formerly required access to working machines. By the end of the following decade the leading mechanics' journal of the time, the highly illustrated

Scientific American, regularly produced drawings and plans of machinery and inventions. The publishers of *Scientific American,* the patent agency of Munn and Company, also reproduced tables of mechanical movements in their guides for inventors.[48] With the aid of such drawings and engravings, inventors and mechanics could visualize how a machine looked in three dimensions and how its parts moved through space. Later, circuit diagrams accompanying works on electrical technology made it possible to follow the flow of electricity through the sections of a circuit, much as a drawing made it possible to follow the movement of the interconnected parts of a machine.

The Social Process of Nineteenth-Century Invention

Because the patent system rewarded only the original inventor, it encouraged inventors to present their own work in terms of an initial inventive insight, with the rest of the process a mere working out of the idea. The process of invention itself has encouraged such views. As Eugene Ferguson and Brooke Hindle have demonstrated, not only were visual images important in transmitting information about machines, but technological design itself generally required a visual, spatial mode of thinking. The beginning of the inventive process was the imaginative conception of a design in which a particular combination of component parts was arranged into a new pattern. This imaginative construction is that flash of inventive insight that is commonly considered the actual creative act. We have come to view such inventive insights as the key element in nineteenth-century invention largely because of the public descriptions of their work offered by inventors seeking recognition and reward. Yet the act of creating a useful invention encompassed much more.

Even before beginning to design an invention, inventors acquired knowledge through both practical experience and reading scientific and technical literature. Although nineteenth-century invention is usually regarded as highly empirical, technical journals constantly emphasized how important it was for inventors to grasp the basic scientific principles of mechanisms and processes and thus avoid wasting time and money on inadequate designs. Indeed, scientists such as Joseph Henry argued that true invention was in fact the discovery of general principles. Most Americans saw technological invention in a more practical light. Because they believed that general principles only created something of value for the society when applied to the design of particular mechanisms, Americans designed their patent law to exclude such principles from patentability.[49]

Furthermore, Americans involved in developing new technology recognized that the laws of mechanics and other scientific principles provided only one source of knowledge and inspiration to inventors. More significant, particularly for the majority of inventors whose inventions were improvements on existing machines and techniques, was awareness of the existing state of the art. It was for this reason that patent and other models of machinery were considered a crucial aid and inspiration to inventors. Furthermore, it was asserted that through the study of new technology the "inventive faculties . . . become excited and urged on by a spirit of emulation: invention follows invention."[50] This spirit of emulation produced an ever-increasing number of inventions. As the development of new industries created a fraternity of operators and mechanics familiar with specialized machinery, their experience often led to minor but important modifications. While not all such improvements were patented, inventors and their advocates argued that novelty and utility could involve little more than "the addition of a screw, or of a peg" as long as it "renders that valuable which was of little comparative worth."[51] The proliferation of American patenting activity by the time of the Civil War was spurred by an expanding industrial economy, but it was also a direct outgrowth of the successful liberalization of the patent examination system caused by agitation on the part of inventors and their advocates who promoted this view of patentable inventions.[52]

Ultimately, inventions needed to actually perform the tasks for which they were conceived and inventors found a crucial part of the inventive process to be the actual development of a working device. This was an interactive process of construction, experimentation, and redesign, involving not only materials and mechanisms but often the talents of skilled mechanics as well. Although the literature of invention recognized the need for such experimentation, it failed to give credit to the skilled craftsman who constructed devices and whose own design ideas, growing out of familiarity with materials and mechanisms, often modified an inventors' design in subtle ways that helped to make it successful. Although accounts of successful invention often recognized assistants and occasionally even more obscure mechanics, the patent system and popular attitudes regarding the nature of inventive insight acknowledged only the originator of the idea as the true inventor. The contributions of those who played an important role in turning the idea into a practical device were generally ignored. Contemporaries and historians alike have generally overlooked the cooperative nature of nineteenth-century invention.

The individual inventors of the nineteenth century were not the lone inventors of mythology. Too little attention has been paid to the extent to which they were members of technical communities that provided the crucial resources of knowledge and skill, and often of employment and financing as well. The cooperative nature of technology was occasionally noted and implicitly recognized through the efforts of mechanics' institutes to aid inventors, and through attempts to form inventors' institutes. Such formal cooperative organizations provided limited aid to inventors; their existence made more explicit the tension between the myth of the heroic lone inventor and the realities of inventive activity. More common and ultimately more helpful to the inventor were less formal communities formed by those engaged in operating or producing machines for various industries. Industrial specialization created invaluable knowledge that circulated informally among those actively engaged in an industry. Later in the century, technical journals devoted to specific industries opened more formal avenues of communication. These industrial communities not only exchanged knowledge and the mechanical skills necessary for the technical development of new inventions, but they also helped secure financial backing for experimentation. And at the innovation stage they became the primary arena in which to market inventions.[53]

Innovation and development were distinct but related stages of the inventive process and required inventors to act as entrepreneurs. The entrepreneurial aspects of invention were implicit in a patent system that made potential future remuneration the only reward for the time and expense of experimentation. Although some Americans advocated a patent system that directly encouraged and rewarded inventors, most assumed that significant work would be rewarded in the marketplace.[54] They would have agreed with the statement made by James Whitney during his presidential address to the New York Society of Practical Engineering that "one of the most useful attributes of the patent law is that it leads individuals to furnish means to practically construct, test, and introduce improvements that without such aid would languish and die."[55] Inventors required financial support both at the development stage, when money was needed for experiments, and at the point of innovation when seeking to market a new invention. Although not all inventors possessed entrepreneurial talent or were comfortable in such a role, they could find individuals with the necessary talent and connections to assume this crucial function, which could mean the difference between failure and success.

The entrepreneurial context of American invention was fostered by social values favoring private development of the nation's resources and by economic opportunities made possible by canals and railroads that expanded markets and encouraged manufacturers to expand their operations. Because new technology facilitated their efforts to expand production, both manufacturers and farmers encouraged invention. The patent system incorporated the values of the marketplace. Because most inventions were individual machines, inventors either sold them outright or organized companies, usually partnerships operating on a small, informal scale, to manufacture and sell the machines themselves. The birth of the telegraph, however, marked a new stage in the industrial evolution of the nation. Rather than individual machines, telegraphy consisted of a system of interrelated parts designed to provide a large-scale, national communications system. For this reason the federal government played a crucial role in its early development and introduction.[56] Federal support was also fostered by an American tradition of mixed enterprise, involving cooperation between the public and private sectors, for internal improvement projects. American social values, though, encouraged private rather than public enterprise, and under private development a few large firms quickly came to dominate the telegraph industry. The development of the telegraph industry by mid-century marked the beginning of an industrial evolution from local, small-scale, informally operated firms to regional and national, large-scale, hierarchically managed corporations. Similar changes were also occurring in the railroad industry, but not until the end of the century did this shift take place throughout the American economy as financial power became concentrated within various industries. Such economic centralization had important implications for those American social values and institutions that supported the work of independent inventors. An examination of the birth and growth of telegraphy reveals the changes taking place in the American system of invention during the course of the nineteenth century as the heroic model of the individual inventor gave way to a modern model of organized industrial research.

2

Invention and the Development of the Telegraph Industry

T
HE INVENTION of the telegraph relied on new knowledge about electricity that emerged in Europe and the United States between 1750 and 1830. Early attempts to use electricity for transmitting information grew directly out of experiments on current electricity undertaken by an international scientific community. However, only as knowledge of electrical science was combined with the practical design knowledge possessed by a group of skilled mechanics in the 1830s and 1840s did electric telegraphs evolve as practical instruments of communication in Europe and the United States.

In the United States, the practical foundations of the telegraph combined with existing ideologies of invention and entrepreneurship to encourage the development of competing telegraph systems. At the same time, the systemic nature of telegraph communication fostered large-scale integration of a national telegraph network that was to stimulate the growth of centralized control. In Europe, this led to nationalization, while in the United States political and economic ideals encouraged private efforts to achieve centralized control and resulted in the creation of the first truly national corporation, the Western Union Telegraph Company.

Inventing the First American Telegraph

Samuel Morse was the central figure in a scientific-technical community, based on specialized knowledge and important social connections, that created the first American telegraph system. While Morse's training as an artist provided conceptual skills similar to those needed for the devising of successful technological improvements, he possessed neither the necessary equipment nor the requisite mechanical skill to construct a telegraph instrument suitable for commercial introduction. He also had only limited knowledge of the recent scientific advances in electromagnetism that made a system of electrical telegraphy possible by the third decade of the nineteenth century.

However, Morse's ability to move between the two cultures of science and technology, to act as what Hugh Aitken termed a "translator," was essential in producing the first American electric telegraph.[1]

Although his grasp of electrical science was incomplete, Morse was one of the few inventors in the country with the economic and social status to acquire the kind of education that gave him ready access to literature dealing with the still little known science. Morse was fortunate in having a father among the professional classes and in achieving such status himself. Only a small percentage of American inventors benefited from such favorable circumstances.[2] The son of a Yale-educated minister, Morse also attended Yale College and later became a professor at the University of the City of New York (later New York University), where he was able to establish connections with the New York scientific community.[3] Morse was probably introduced to scientific ideas by his father even before attending Yale. Jedediah Morse was a minister in Charlestown, Massachusetts, and author of the first American work on geography. It was at Yale between 1807 and 1811 that Samuel developed his initial interest in electrical science. There he attended classes that included lectures and demonstrations on electricity and even assisted with some experiments. Morse renewed his interest in the science in 1827, following a series of lectures by Columbia College professor James Freeman Dana. Subsequently the two men spent a great deal of time together in Morse's New York art studio discussing the subject. Thus, five years later, while returning from an extended stay in Europe, Morse was well prepared to respond to the comments offered by a shipmate, the Boston chemist Dr. Charles T. Jackson, regarding electrical communication.[4]

At dinner one evening, the shipboard conversation turned to André Ampère's experiments on electromagnetism and Jackson noted that the speed of electricity was not retarded by the length of the wire. Morse responded that he saw "no reason why intelligence might not be instantaneously transmitted by electricity to any distance."[5] Fascinated by this notion, Morse spent much of the remainder of his voyage on the *Sully* contemplating methods for accomplishing such an effect. Morse recorded his ideas visually in a series of drawings in a small notebook. These sketches helped him to conceptualize the basic design of instruments, as well as variations on them, in much the same way that he often drew images in his notebooks for future reference in the design of paintings.

Historian Brooke Hindle has demonstrated that conceptual design had similar attributes in both the mechanical and fine arts, requiring visual, spatial thinking and the ability to combine parts into a finished whole. Because of this, a number of early American artists, including Morse, Robert Fulton, and Charles Willson Peale, made contributions to technology.[6] Even before Morse undertook his work in telegraphy, the direct relationship between design in the fine and mechanic arts encouraged him to pursue invention as a potential avenue to financial success. This was true especially at times when his artistic career looked unpromising. Morse undertook his first inventive effort in 1816—a pump for fire engines—after failing to gain financial support for an intended historical painting. The lack of support for such artwork in America forced Morse to focus on commercial portrait painting, though he was never happy with this career. Samuel's brother Sidney joined in the invention of the pump and seems to have had the greater responsibility for actual construction of the device. Samuel apparently was most responsible for the initial conceptualization and for the subsequent drawings that accompanied their successful patent application. Though encouraged by those who examined the pump and declared it to be an important improvement, Samuel and his brother were unable to achieve success in the marketplace, leading Samuel to complain that "an inventor earns his money the hard way."[7] Though his pump did not bring the hoped-for financial success, Morse turned again to invention during another slump in his artistic career. His second invention was a machine for carving marble to produce perfectly copied statues from any model. Morse took his design to Hezekiah Augur, a New Haven mechanic, who constructed a working machine. Augur later developed successful inventions of his own based on such a device, but Morse again failed to achieve financial reward from his inventive endeavors.[8]

The drawings Morse made in his notebook represented a general plan for a telegraph system, but only after arriving in New York could he attempt to translate his ideas into working instruments. His first effort, within a few days of his arrival, was to make saw-toothed type, for use in his transmitter, similar to those he had sketched on shipboard. He cast these while living at his brother Richard's house, and his sister-in-law recalled that "he melted the lead, which he used, over the fire in the grate of my parlor, and, in his operation of casting type, he spilled some of the heated metal upon the drugget, or loose carpeting before the fireplace, and upon a flag-bottomed chair, upon which his mould was placed." However, Morse's continuing efforts to make a career as a painter combined with limited funds prevented him from proceeding further with the invention. Only after accepting an appointment to teach art and design at the University of the City of New York in 1835 did he acquire the financial resources that enabled him to continue his inventive work.

Through his position at the university, Morse not only acquired secure employment for the first time in his career, but he also acquired space in which to experiment. Nonetheless, the conditions under which Morse built his first transmitter and receiver were nearly as crude as those under which he cast his first transmitting type. Morse made these instruments in the small

Figure 2.1. This drawing of Samuel Morse in his studio laboratory and shop at New York University presents a popular image of the lone inventor in his garret. (*Source:* Prime, *Life of Morse,* 289.)

studio at his university apartment that he also used as a laboratory and shop. A drawing of Morse working on his instruments in this room captures the popular image of the lone inventor in his garret struggling to bring his ideas for instantaneous electric communication into practical operation. More significant to Morse than his crude workshop was his access to essential knowledge and skills possessed by others in the university community. His first important association was with Leonard Gale, professor of chemistry, geology, and mineralogy at the University. In a period when electricity was closely allied with chemistry, Gale was well aware of recent electrochemical and electromagnetic research.

Though the earliest attempts to devise an electric telegraph date from the second half of the eighteenth century, practical telegraphy depended on the scientific development of electrochemical batteries during the early nineteenth century. The earliest telegraphs were built in Europe and used a current produced by common scientific apparatus for generating frictional electricity. Occasionally a Leyden jar was used to store and discharge this static electricity. A crucial step toward practical telegraphy was taken by Allesandro Volta and others whose important electrochemical experiments led to the development of electrochemical batteries to provide a steady current. Morse would certainly have been aware of many of these experiments from his studies at Yale and his discussions with Dana, but Gale had a much more intimate familiarity with the most recent research in electrochemistry, as well as practical experimental experience.

Gale also possessed the critical knowledge of recent work on electromagnetism, including the important work of Joseph Henry, which made an electromagnetic telegraph practical. Gale knew Henry and had, in fact, performed a series of experiments with him in 1833. He was also aware of Henry's scientific writings on electromagnetism and pointed these out to Morse, who had only slight knowledge of the work of Ampere and others experimenting with electromagnetism.[9] Although the exact nature of Henry's contribution to the telegraph became the focus of a dispute that persists to the present day, his work and that of other researchers investigating electromagnetic phenomenon played a crucial role in the successful development of Morse's invention.[10] Morse doubtless had gained some familiarity with electromagnets from his conversations with Dana, who had constructed an electromagnet, on the plan of its inventor William Sturgeon, for his 1827 lectures. Furthermore, Morse's conversations with Charles Jackson on board the *Sully* in 1832 focused on electromagnetism. Yet it was Gale who provided the crucial knowledge of Henry's experiments on electromagnets that made their adoption for practical telegraphic purposes possible.

Joseph Henry made two critical contributions to the study of electromagnets that were incorporated into Morse's telegraph system. First, he devised more powerful electromagnets in two basic designs. One consisted of a single insulated wire wrapped several times in multiple layers around an iron core and the other used a number of smaller windings of insulated wires coiled separately around the core. Second, he determined that each magnet design required different battery arrangements. The single-coil magnet was most effective in circuit with a battery of several cells wired in series to achieve higher voltage, while the multicoil magnet worked best with either a single-

Figure 2.2. In his 1831–32 experiments to determine the relationship among batteries, conductors, and electromagnets necessary for producing magnetic power at a distance, American scientist Joseph Henry demonstrated a telegraph when he magnetized a piece of iron and made it produce an acoustic signal. (*Source:* F. L. Pope, "American Inventors," 932.)

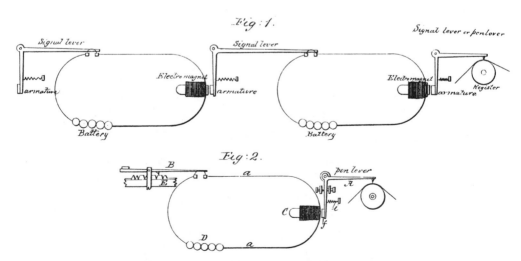

Figure 2.3. Morse's relay design (U.S. Pat. 1,647), which incorporated information derived from Henry's experiments, used an electromagnet to bring a local battery into the circuit in order to provide sufficient current to activate the register.

cell battery or one with several cells wired in parallel to achieve higher current. Henry's research provided the theoretical basis for a key element of Morse's system, the electromagnetic relay, which opened and closed a local circuit with its own battery to work the receiving instrument. Though the origins of this device remain in dispute, it is probable that through working with Gale, Morse drew indirectly on Henry's work in conceiving it.

Regardless of the origins of the relay, Gale's experimental and theoretical knowledge of batteries and electromagnets, especially his knowledge of Henry's work defining the relationship between them, was crucial to the design of the electrical components of Morse's system. Morse recognized the importance of Gale's contribution by making him a partner and asking him to "devote his professional services and skill . . . in effecting improvements in the philosophical and physical qualities or properties of said invention, and in extending and perfecting the same by such chemical and scientific experiments as may be suggested as serviceable and proper."[11]

While Morse's telegraph was powered by electricity and relied directly on advances in electrical science for its origins, the transmitter and receiver were essentially mechanical instruments, which required traditional mechanical skills for their design. Though Morse's training as a painter refined conceptual skills very similar to those needed for mechanical invention—each activity called for spatial, visual thinking and the ability to combine parts into a finished whole—Morse possessed neither the equipment nor mechanical skill necessary to construct instruments suitable for commercial introduction. The American machine-shop process of invention and development thus played a crucial role in the development of the telegraph.

Morse had relied on his practical experience in designing his first instruments. His receiver design incorporated the familiar wooden canvas stretcher he used as a painter. At the center of a wooden bar placed across the stretcher he attached an electromagnet that, when electrified, attracted an iron lever suspended from the top of the stretcher frame. This caused a pencil to move across a roll of paper, which was drawn over a roller at the bottom of the framework by the mechanism of an old clockwork. Morse derived his transmitter from the printer's type and composing stick he was familiar with from his brother Sidney's publishing activities.[12] Into the composing stick he placed type he had cast himself with coded teeth on the bottom. A hand crank moved a lever with an electrical contact on its tip over the teeth, thus making and breaking the circuit and transmitting a signal over the line. These crude instruments, handmade in his university studio from readily available materials, enabled Morse to demonstrate the basic principles of his invention, but they were inadequate for practical use. It required a trained mechanic to take Morse's designs and produce instruments suitable for public demonstrations and potential commercial introduction.

Figure 2.4. Morse's original telegraph instruments were influenced by his practical experience. The familiar wooden canvas stretcher he used as a painter was converted into the transmitter, while the printer's portrule and type familiar from his brother's publishing enterprises were incorporated into the transmitter. (*Source:* Reid, *Telegraph in America,* 54.)

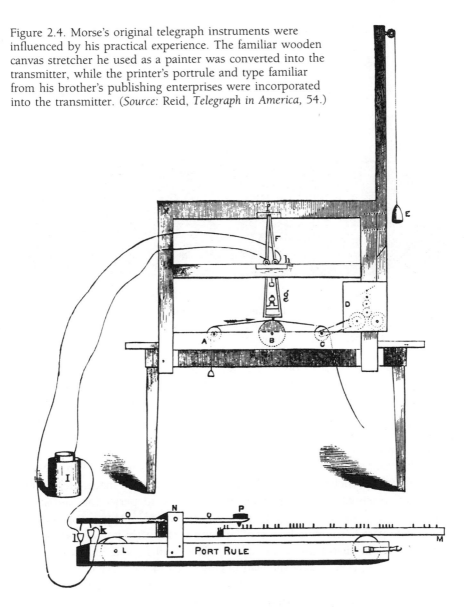

As an artist and professor Morse might have had limited access to the world of the mechanic. However, his professorship at the university brought him into contact with Alfred Vail, who had received a practical education in the machine shop of his father Stephen's Speedwell Iron Works. Alfred's decision to study at the University of the City of New York, rather than follow his father into manufacturing, brought him into contact with Morse. Vail was not only a member of the wider mechanical community, but one

Figure 2.5. The Speedwell Iron Works, shown in the company's billhead, provided Morse with access to the machine-shop culture of nineteenth-century invention which proved crucial to the practical development of the electric telegraph. (*Source:* F. L. Pope, "American Inventors," 928.)

whose educational background prepared him to work with the new motive force of electricity. In September 1837, Morse entered into a partnership with his former student, which provided Morse with the resources of Vail's mechanical skill and of his father's Speedwell Iron Works. In the two years before the Vails became involved in his invention, Morse's progress was slow, but by January 1838 he was able to publicly demonstrate the system.

The exact nature of the work done at Speedwell is unclear. While Morse undoubtedly participated in some of the design changes made there, Vail's help was essential "in constructing and bringing to perfection, as also in improving the mechanical parts of said invention."[13] Morse was only at Speedwell part of the time between September and early January while construction of the instruments was taking place, and he was ill during much of his stay. Vail clearly took the lead in construction and design work on the new instruments. The importance of Alfred Vail's mechanical know-how in constructing a practical instrument is suggested by his letter to Morse of 14 October 1837, in which he states, "I have dispensed with the large spiral wheel, and have concluded to use one smaller on the opposite end of the shaft. The reason for doing so, was that three inches was not more than enough for the magnet, and that the large spiral would be in the way of the magnet. Now, the space between the drums will be clear and free. I think you will like the last plan best."[14] The register discussed by Vail in this letter

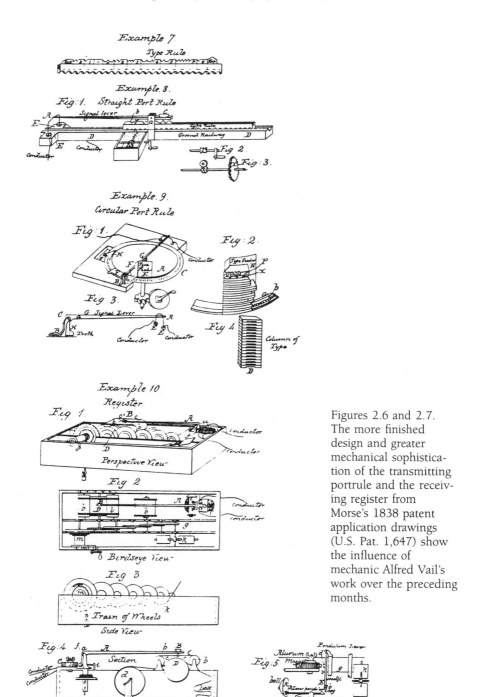

Figures 2.6 and 2.7. The more finished design and greater mechanical sophistication of the transmitting portrule and the receiving register from Morse's 1838 patent application drawings (U.S. Pat. 1,647) show the influence of mechanic Alfred Vail's work over the preceding months.

was very different from Morse's original canvas stretcher design as evidenced by the patent drawings executed a few months later (U.S. Pat. 1,647).

Over the next several years, Vail seems to have made many contributions to the design of telegraph instruments that were publicly credited to Morse. Two reasons suggest themselves for Vail's failure to claim any of these designs himself. The first was his understanding regarding a clause in Vail's original agreement with Morse. When years later Vail privately asserted his claim to an important modification of the register, he noted that any inventions made by he, Gale, or F.O.J. Smith, the chairman of the House Commerce Committee who acquired an interest in the Morse patents in 1838, "would belong to all jointly, and become part of the original invention. I could not therefore have taken out a patent for this invention myself."[15]

A second reason is suggested by Vail's letter to Morse of 19 March 1838,[16] in which he not only described his progress on Morse's instruments, but also his own work on the printing telegraph he had devised the previous September. In the letter Vail seems to make a distinction between improvements he made to Morse's original design and those he felt were wholly original. Vail described his work on a new port rule in terms that suggest to the reader real inventive work: "I have contrived a new port rule . . . which will not require sticks or rather but one stick. It was brought forth after much study and perplexity. The port rule is a difficult affair. I shall make it run by machinery." Yet, in discussing his printing telegraph Vail makes a distinction between Morse's telegraph and his own. He is careful to suggest that bringing his invention before the public "would not in the least hinder the progress of the Telegraph. Both might be attended to with care, as the Printing Machine will require no experimenting. I would not for a moment propose it, did I think our interest would be neglected or slighted." This distinction between refining Morse's telegraph and developing his own design is also evident in Vail's book on the telegraph where he staked a claim to the printing telegraph while only describing and not attributing the various port rule designs.[17]

By 1844, Vail and Morse had substantially redesigned their initially cumbersome instruments. Those used on the first experimental line were significantly more simple and elegant in their mechanical construction.[18] In their modified form, the transmitting and receiving instruments were extraordinarily uncomplicated. The transmitter was a simple key, which opened and closed the circuit, thereby transmitting a sequence of dots and dashes that made up the Morse code. These signals were then recorded by a register that was similar in design to the old canvas stretcher receiver, but which was mechanically more refined. Vail was responsible for most of the mechanical refinement of the register and has been credited with the invention of the key. He was also responsible for later commercial development of the instruments,

Figure 2.8. By 1844 Vail had simplified the mechanical design of the receiving register by replacing its complicated train of wheels with a more simple clockwork mechanism. (*Source:* Smithsonian Institution photo 29651.)

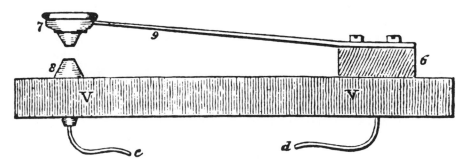

Figure 2.9. Vail altered the transmitter by isolating the signal lever of the portrule design and converting it into a simple make-and-break contact key. (*Source:* Vail, *Electro Magnetic Telegraph,* 22, 33.)

Figure 2.10. The original operator's station of 1844 included the register on the left with a clockwork drive to move the paper from the reel on the right under the recording lever activated by the electromagnet. Next to the paper reel is the crude early transmitting key. (*Source:* Vail, *Electro Magnetic Telegraph,* 19.)

which also owed much to the mechanical skill of other machinists who produced the instruments used in 1844 as well as those used in 1846 on the first commercial lines.[19]

Morse was clearly the central figure in the development of an American telegraph system, but a variety of other individuals contributed to its evolution between Morse's initial conception in his 1832 drawings and its first commercial introduction in 1844. The difficulty in determining the exact contributions of those associated with Morse stems in part from the traditional heroic model of invention. Because the patent system rewarded only the original inventor, most people identify the originator of an idea as its sole inventor. This is particularly true of any invention that produces a new technology. Thus, Morse is popularly credited with the invention of the telegraph because he conceived of its essential elements. Joseph Henry was responsible for a critical element of the invention, but not for its application to telegraphy and Morse appropriately received credit for the application. Furthermore, the two men's motivations were different. Henry used his work on electromagnets to extend scientific theory, while Morse focused on their practical, commercial use. Yet, Alfred Vail and Leonard Gale also worked with Morse toward practical applications of telegraphy, but their contributions have been generally viewed as those of assistants helping to bring the invention into final form rather than as contributors to the basic design. Vail in particular is often thought of as a constructor and not as an inventor, yet both the transmitting key and register actually used in the system appear to have been largely his design. Furthermore, even the code named after Morse, which enabled words to be translated into electric signals, may well have been revised into its common form by Vail.[20] To credit Morse as the inventor of the first American electric telegraph is to ignore the collaborative character of that invention.

Nonetheless, Morse did play a central role in the creation of the first American system. The development of the system relied on his abilities to envision the elements necessary for a telegraph system, to gain access to essential skills and knowledge, and to maintain his commitment to the invention in the face of severe trials. His role was much like that of a team research leader. Such a concept, however, would have been foreign to Morse and his contemporaries. Instead they conceived of him simply as the inventor. Such a conception fit well the heroic model embodied in the patent system, which emphasized the original flash of insight while portraying the development of a working device as the mere reduction to practice of the original idea.

Introducing the New Invention

Reducing the invention to practice enabled Morse to acquire patents that conferred a temporary monopoly for his idea. However, patents only provided proof of ownership. Selling the idea in order to acquire the financial support needed to bring the new telegraph into public use and to gain what Morse considered his proper financial reward for the sacrifice and work that had gone into its development proved particularly difficult in the case of the electric telegraph. In the 1830s resources available for the development of such a system were limited. Most inventions consisted of individual machines that their developers generally sought to manufacture and sell or to license for others to make and sell. But the manufacture of telegraph instruments would return few profits until the entire system came into common use. One of the early technical analyses of the new invention suggests another important reason why private development capital might have been difficult to obtain. In February 1838, the Committee of Science and the Arts of the Franklin Institute reported favorably on the new invention but cautioned that "any effectual plan [for distributing the wires] must be very expensive."[21] Consequently, Morse envisioned the telegraph as a form of internal improvement like the canals and railroads, and he turned to the national government as the only institution with sufficient capital resources for such a project.

Although federally funded internal improvement programs declined in popularity during the 1830s, specific programs still received support and

Morse had good reason to believe that Congress would finance his telegraph. Early in 1837 Congress received petitions calling for the establishment of a mechanical system of semaphore telegraphs. Samuel Reid, proprietor of a such a system of semaphore in New York Harbor, submitted a petition to Congress in which he proposed building a line of such telegraphs from New York to New Orleans via Washington. Another proposal came at the same time from John Parker of Boston, the inventor of another system of semaphore telegraphs, who urged Congress to establish telegraph lines throughout the nation and solicited the support of the postmaster general for his plan.[22] Congress responded by requesting that the secretary of the treasury initiate an official investigation to "determine the propriety of establishing a system of telegraphs for the United States."[23]

The reasons for Congress's interest in a telegraph system remain obscure, but the potential military applications of such a system and its usefulness as part of the federal postal system probably provided the greatest impetus. Champions of a national telegraph system also addressed the general problem of intercommunication in the expanding nation by suggesting that it would benefit the nation's commercial interests as well.

Congressional support was vitally important to Morse, who used it to seek assistance from the Vails. He signed his partnership agreement with Alfred four days before writing to the secretary of state concerning his own telegraph system.[24] Morse clearly believed that his electric telegraph system presented important advantages over others submitted to the attention of the secretary of state and that these advantages would enable him to acquire congressional funding. Stephen Vail's willingness to invest in his son's and Morse's work on the electric telegraph derived in large measure from a belief that government adoption of the invention would return him a handsome profit on his investment.

In his letter to the secretary, Morse was able to make a strong case for the technical superiority of his electric telegraph. The key difference between Morse's system and others proposed to Congress was his mode of transmission. All of the other proposed systems were for semaphore telegraphs, consisting of a series of towers, located every few miles, and mounted with signaling devices that transmitted in code to observers at each tower equipped with telescopes.[25] The proponents of semaphore telegraphs claimed they could transmit messages between New York and Washington in about twenty-five minutes, and to New Orleans in less than two hours. Morse, on the other hand, promised instantaneous communication. Furthermore, his electric telegraph continued to operate in conditions of poor visibility, an important consideration for the government officials investigating the various telegraph systems.[26]

During the next few months, Morse successfully demonstrated his invention before public audiences, most notably the Franklin Institute, which issued its favorable report, and in April 1838 the House Commerce Committee recommended an appropriation of $30,000 to build an experimental telegraph line using Morse's system.[27] But the pressure of dealing with the economic repercussions of the Panic of 1837, as well as the pressure of other legislative concerns, prevented Congress from approving the appropriation. Nonetheless, Morse continued his lobbying efforts, while at the same time seeking private investors. The effects of the panic, as well as Morse's determination that control of the invention would revert to the federal government if Congress decided to purchase it, made it difficult to acquire private support. Morse therefore renewed his congressional lobbying effort and obtained another favorable report from the House Commerce Committee in December 1842.[28]

In lobbying Congress for support, Morse argued that successful operation of telegraph lines in European countries was preventing the world from recognizing American claims to the invention. The European telegraphs were undoubtedly a compelling argument in favor of congressional support, but not because of these nationalistic appeals, with which Morse had been attempting to stir interest for many years. Instead the European telegraphs provided demonstrable proof of the practicability and usefulness of the electric telegraph. More important to the debate was concern that private ownership of the telegraph would produce "serious injury" to the Post Office department, whose monopoly over postal communication was generally considered an appropriate government function. However, to help insure passage the bill was amended to place the appropriation under the control of the secretary of the treasury rather than the postmaster general, who opposed "Morse's foolishness."[29]

Finally, on 3 February 1843, Congress passed the committee's bill "to test the Practicability of Establishing a System of Electro Magnetic Telegraphs in the United States," authorizing $30,000 for Morse to build an experimental line. The passage of the bill was aided by improved financial circumstances, as well as a Whig government more favorable to internal improvements. The debate and vote on the bill suggest that the opposition to it was based largely on two issues—the experimental nature of the telegraph and sectional opposition to internal improvement measures. The first objection proved relatively unimportant as Morse and his supporters could point to successful European telegraphs. Only twenty-two members voted for Congressman Cave Johnson's scornful amendment calling for half the sum to be spent for experiments on mesmerism as another magnetic science worthy of study. The close vote (90 yes, 82 no, and 70 abstentions) on the appropriation bill itself, however, was largely cast along sectional lines, and the Whig majority in the House was crucial to its passage.[30]

Congress's appropriation provided the rather substantial development capital that was difficult for the untried system to obtain from private entrepreneurs. With this funding Morse and his associates built the first experimental line between Washington and Baltimore. Before this line was placed in practical operation, Morse's telegraph system required additional refinement. Alfred Vail, Leonard Gale, and James Fisher, a professor of chemistry at the University of the City of New York, assisted in the building of the line and the further development of the system. Vail, who served as Morse's primary assistant, continued to make important improvements in the mechanical design of the register, superintended the manufacture of instruments, and conducted experiments to determine the appropriate battery power for the most efficient operation of the line. Among Vail's most important contributions during the early years of operation was the introduction of the simple key transmitter, which became the standard signaling device.[31]

Although Morse and his associates also conducted careful electrical experiments and compared their results with those derived from the electrical laws of scientists such as Georg Ohm and Heinrich Lenz, they were hampered by the lack of reliable practical experience with electric telegraphs.[32] Thus, on the basis of reports indicating the successful use of underground lines in England, Morse chose this mode of installation for his own lines, believing that underground wires would be better protected and more durable than wires strung overhead on poles. James Fisher was placed in charge of the wire manufacture and superintended its placement in the underground tubes. At the behest of F.O.J. Smith, one of the assignees of the Morse patents, mechanic and inventor Ezra Cornell devised a trenching machine to lay the lines. Poor insulation and other technical difficulties beset the underground line. Although Morse blamed Fisher for not properly inspecting the pipes and discovering the insulation problems, Morse appears to have had little understanding of the problems posed by underground lines and was himself largely responsible for the failure of the experiment. The tremendous expense involved in installing the underground line placed the entire project in jeopardy and Morse fired Fisher, ostensibly to save the money from his salary. He also had Cornell break the trenching machine in order to gain time while seeking a solution.[33]

Cornell and Vail sought the solution to this problem in the literature on European telegraphy available at the Library of Congress and the Patent Office. In his reading, Vail discovered that in fact problems had been experienced with underground lines in England, causing the wires to be strung overhead on poles instead. Both Vail and Cornell proposed plans for overhead lines, and after consultation with Joseph Henry, Morse chose Cornell's plan to use glass drawerknobs to insulate the wire from the wooden poles, thereby preventing current from escaping the line and disrupting transmission. Overhead lines—less costly and quicker to build than underground lines—were well suited to the vast expanse of the American continent. Even if the underground line experiment had proved successful, it is unlikely such lines would have been used outside of the major cities. Construction of the overhead line between Baltimore and Washington took less than two months and Morse transmitted the first message on 24 May 1844.[34]

Although this line demonstrated the practicality of the new technology, it did not prove profitable over the next two years and the government refused to adopt the telegraph as a public enterprise.[35] The government's failure to purchase Morse's invention is not surprising. By the mid-1840s American development policy had turned away from direct public funding of internal improvement projects. This shift was produced by Jacksonian attacks on special privilege and their attempts to restrict the power of government, increase liberty, and "enlarge the options open to private individual and group energy."[36] The Panic of 1837 and the ensuing years of economic hardship also encouraged this change in policy as state debts grew due to overambitious financing of internal improvements. In order to encourage mobilization of capital for such improvements, states began passing more liberal incorporation laws that reduced financial risks for individual investors and stimulated the purchase of stock in corporations that became the favored vehicle of American economic development. Although private corporations took responsibility for internal improvement projects, government financing continued to be important. Indeed, Carter Goodrich has described American development programs as mixed enterprises. Although most Americans favored decentralized government and opposed national planning of internal improvements, many also supported some government funding of such projects, but emphasized private leadership in joint undertakings as best serving the interests of individual free enterprise.[37]

Morse's own views reflected the complexity of American thinking regarding the funding of internal improvements. Morse had grown up in a Federalist stronghold in Massachusetts and, although he had become a Democrat, his economic views betrayed his upbringing. He was a greater advocate of government aid to special privilege than most Democrats, although he believed such privilege should be based on accomplishments. As a painter he had sought government support for his historical paintings, and as an inventor he believed that he should receive a lifetime government pension like that recently awarded by the French government to Louis Daguerre for his photographic process.[38] Morse's hopes for a similar award from the American government recalled earlier calls for direct government subsidies. Most Americans, however, had come to view government premiums as undemocratic, believing that they prevented equal competition and inhibited rather than fostered invention, which was linked with private enterprise development policies.[39]

In the case of telegraphy, development through private enterprise did encourage invention. The shift from government control to commercial development of the Morse telegraph encouraged private entrepreneurs to support the efforts of other inventors to develop competing systems. These inventors had before them the example of a successful telegraph invention and the personal model of Morse as a heroic inventor. In a society that believed that individual economic success was a primary social goal, the heroic inventor could inspire others as a prominent example of the self-made man.

Americans came to exalt the self-made man as a national ideal as economic and political changes during the Jacksonian era helped to consolidate liberal capitalism as the dominant ideology in American society.[40] Classical republicanism had been conceived and promulgated in an agrarian society basically hierarchical and deferential in character, though Jefferson and others sought to extend political leadership to a natural aristocracy based on talent and virtue. Under the pressures exerted by participation in a commercial capitalist economy and the seemingly unlimited opportunities offered by abundant resources, Americans replaced the older organic vision of society contained in classical republicanism with a dynamic new vision. National progress was envisioned as the result of individual success achieved by embracing economic opportunity. Instead of following a given calling, the enterprising self-made man realized success by finding the "likeliest vehicle for personal distinction."[41] Because invention was associated with personal fame and fortune but also perceived as a way of bringing practical benefits to the nation, it became a likely "vehicle for personal distinction."

Though embracing enterprise and wealth as signs of personal success, many Americans remained ambivalent about the naked pursuit of wealth. For this reason such ambitions were often placed in a moral context. This moral ambivalence toward a success ethic founded on wealth also affected the myth of the heroic inventor. Thus, even those actively promoting invention as a means of self-advancement claimed that the inventor's "real recompense . . . is the consciousness of doing one's duty."[42] Many American inventors professed "a deep interest in the welfare of [their] fellowmen."[43] Morse echoed this view in an 1837 letter to his close friend, Catherine Pattison, when he confided that

> The condition of an Inventor is indeed not enviable. I know of but one consideration that renders it in any degree tolerable and that is the reflection that *his fellow man may be benefitted by his discoveries*. . . . Many are ready to snatch the prize or at least claim a share so soon as the

success of an invention seems certain, and honor and profit alone remain to be obtained. . . . unless there is a benevolent consideration in
our discoveries, one which enables us to rejoice that others are bene-
fitted even though we should suffer loss, our happiness from any
honor awarded to a successful invention is exposed to constant danger
from the designs of the unprincipled.[44]

Nonetheless, Morse became one of the primary examples of the inventor
who achieved both fame and fortune.

Invention and Enterprise

To promote private development of the telegraph Morse turned to former
postmaster general Amos Kendall, who became agent for three-fourths of the
patent right. One fourth was retained by former congressman F.O.J. Smith,
who had become interested in the Morse telegraph after serving as chairman
of the House Commerce Committee that issued the favorable 1842 report on
the invention.[45] Kendall pursued a strategy of licensing regional companies to
construct lines between major cities while retaining the option of selling out
to the government.[46] Lines were built between New York and Philadelphia,
Boston, and Buffalo. The New York–Philadelphia contract went to Kendall,
while Smith received the contract to Boston, and the Buffalo line was contracted to Theodore Faxton, manager of a stagecoach company, and Ezra
Cornell, who had assisted in the construction of the original Washington-
Baltimore line. Kendall also envisioned a western line and contracted with
newspaperman Henry O'Rielly to build one linking his New York–Philadelphia line to St. Louis. These promoters gained profits as stockholders
and as contractors for the construction of lines. They usually raised money
from a host of small investors who were often more interested in the benefit
derived from having access to the telegraph than in the potential profits of
their telegraph companies.

The organization of the new telegraph companies was a product of the
same economic policies that promoted private over public development. As
state governments turned from subsidizing internal improvements to fostering private development of their economies, they passed legislation that
encouraged private entrepreneurs to mobilize large amounts of private capital
for transportation, manufacturing, banking, and other economic activities.
Between 1820 and 1850, state legislatures increasingly granted corporate
charters in a routine and standardized manner. By 1850, they had begun
passing general incorporation acts, thus making the benefits of incorporation,
formerly granted only as a special privilege, widely available.[47]

The limited liability corporation reduced both the risks to individual
stockholders and their direct involvement in the daily operations of the

enterprise. Each stockholder risked only the amount of his individual investment and was not responsible for the debts of the firm as was the case in
traditional family-owned businesses and limited partnerships. Furthermore,
the corporate charter placed duly elected company officials in charge of the
daily operations of the company. These officers were held responsible to the
stockholders in much the same way that government officeholders were
considered responsible to the electorate. While some Americans questioned
the creation of institutions that might consolidate and abuse economic power,
most perceived the ability of corporations to use their larger assets to increase
the production and dispersion of goods as socially desirable, particularly as
it helped diffuse material progress throughout the nation.[48] Although the
telegraph corporation would become a symbol of the abuse of private economic power after the Civil War, the early telegraph companies retained a
democratic character. Perceiving the corporation as a democratic institution,
the first telegraph entrepreneurs organized the industry as a confederation
of allied state-chartered companies. Amos Kendall, in establishing the first
telegraph companies using the Morse patents, sought to create a unified
system on the basis of a common interest in the Morse patent. When Henry
O'Rielly broke with the Morse interests he built a competing system of lines
organized so that the confederation of companies decided on joint policy
democratically through the votes of their representatives on a general board.
The failure of these alliances to effectively coordinate telegraph traffic would
open the way toward greater consolidation in the industry.[49]

Henry O'Rielly's break with the Morse interests, precipitated by a dispute
over the terms of his contract,[50] hastened the introduction of competing
telegraph inventions as he sought to continue operating his lines with alternative telegraph technology. As negotiations between O'Rielly and the Morse
interests broke down in the fall of 1846, he acquired an interest in several
new telegraph inventions.

The first of these was a printing telegraph system patented by New York
inventor Royal E. House. Though little is known of his early life, House
apparently had a fondness for mechanisms and an ability for mentally conceiving machinery. He devised his first invention, a machine for sawing barrel
staves, in 1839 at the age of twenty-five. This invention did not provide
significant financial reward and he left home in 1840 to study law. House
nevertheless continued to be attracted to invention as a vehicle by which he
could achieve distinction. Engaged by the subject of electricity while reading
books on natural philosophy early in his law studies, House became aware
of the work by Morse and others on telegraphy and returned to invention
by working on his own design. Through his reading, House came into contact
with the same scientific community that Morse had drawn on for information
about electricity. In order to build his telegraph House also had to draw on

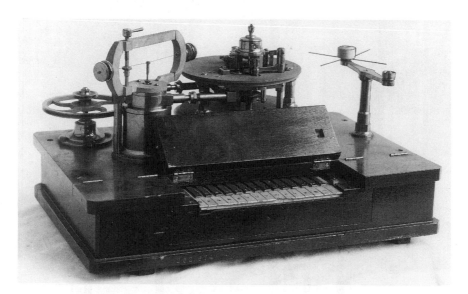

Figure 2.11. This photograph of the patent model of Royal E. House's printing telegraph shows its sophisticated but complex mechanical design. (*Source:* Smithsonian Institution photo 30396).

the mechanical community, and by 1844 he ventured to New York where he found skilled mechanics to construct the parts of his new invention. House's printing telegraph differed significantly from Morse's. Unlike the Morse system, which recorded messages in a code that had to be deciphered, House designed his telegraph to print messages in roman letters. His telegraph incorporated a transmitting keyboard, with individual keys for each letter, which sent electrical impulses to a receiving instrument. There the impulses activated a typewheel to print messages on paper strips. These messages could be instantly delivered to the recipient without further processing.[51]

House found financial support for his inventive efforts from O'Rielly, Hugh Downing, a wire manufacturer and president of the Atlantic and Ohio Telegraph Company, and Samuel Selden, an incorporator of the Atlantic, Lake, and Mississippi Telegraph Company of which his brother was president. However, O'Rielly did not join them in purchasing rights to House's telegraph outside of the areas covered by the O'Rielly lines. When tests of the instrument proved only partially successful, O'Rielly's enthusiasm for the instrument cooled. Downing and Selden, however, used the House system to compete with the Morse lines along the eastern seaboard. Although the lines using the House system provided ineffectual competition for the Morse lines, House's printing telegraph subsequently proved an important competitive tool.[52]

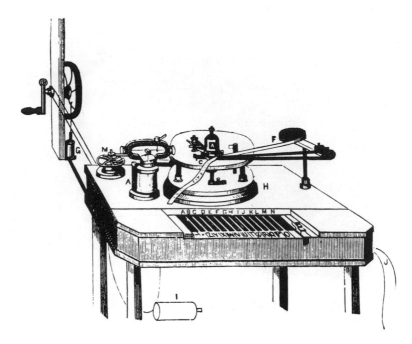

Figure 2.12. One of several diagrams used by Taliaferro Shaffner in his 1859 telegraph book to describe the complex operation of House's telegraph, which required two operators. One of the operators turned a hand-crank that powered the machine, while the other operated the keyboard. A revolving circuit-wheel on the transmitting machine had twenty-eight cogs that rapidly opened and closed the circuit until stopped momentarily by the depression of a key. The signals transmitted by the circuit-wheel caused a series of movable magnets in the receiving instrument to open and close the valves working an air-powered piston that controlled the typewheel escapement and printing mechanism. (*Source:* Shaffner, *Telegraph Manual,* 392.)

O'Rielly himself turned to alternative systems that were little more than temporary expediencies resorted to in order to keep his lines operating. He acquired interests in a telegraph devised by Charles B. Moss, but this instrument, which was never patented, proved worthless. A register devised by Francis Pease was also never introduced into use. More successful was the "Columbian" telegraph invented by operators Samuel Zook and Edward Barnes. O'Rielly used this invention, although it clearly infringed on the Morse patents, until prevented by an injunction in September 1848. However, O'Rielly soon acquired a more important alternative system that appeared not to infringe on the Morse patent.[53]

The design and development of this new system illustrates the attraction of invention as an avenue to financial success within the context of the

Anglo-American ideology of self-improvement. Though a poor student in school, Alexander Bain showed talent as an apprentice clockmaker in Wick, Scotland. While serving his apprenticeship, he attended a lecture on "Heat, Sound, and Electricity" at the Masonic Hall in the neighboring town of Thurso. This lecture stimulated him to conduct his own primitive experiments in electricity. Equally importantly, it caused Bain to quit his apprenticeship and start on a program of self-improvement that eventually led him first to Edinburgh, and then, by 1837, to London. While employed as a clockmaker in London, Bain attended lectures at the Adelaide Gallery and the Polytechnic Institution, and began reading scientific works on electricity. As a result of the new knowledge he acquired about electricity and his own experiments on the subject, Bain devised an electric clock and also conceived an electric telegraph system in 1840. In his efforts to promote these inventions Bain sought the advice of the assistant editor of the *Mechanics' Magazine,* one of the foremost journals of mechanical technology and promoters of self-improvement in Great Britain. The editor suggested that Bain take his inventions to Charles Wheatstone, professor of experimental physics at King's College, who was developing his own telegraph system with William Cooke. The results of Bain's contact with Wheatstone have been the subject of some dispute. Bain later claimed that Wheatstone first offered his assistance and then sought to appropriate the clockmaker's electric clock and electric telegraph inventions. Bain failed to achieve what he regarded as either adequate recognition or recompense for his invention and thus came to the United States in 1848, to seek his fame and fortune by patenting and promoting his telegraph system.[54]

Bain's telegraph differed significantly from both Morse's and House's. It transmitted messages by means of automatic machinery run by clockwork derived from his previous experience as a clockmaker. Operators used specially designed perforating machines to punch holes representing Morse or other telegraph codes onto a strip of paper. They then fed this strip into a transmitter, which passed an electrical contact over the perforations, causing the circuit to close intermittently and transmit signals. A receiver at the other end of the line recorded the signal on chemically treated paper by electrical decomposition. The Bain system often worked better over the poorly insulated lines characteristic of early American telegraphy. It could also apparently be worked over longer distances without relays and with less battery power.[55] Like House, Bain found ready support from entrepreneurs seeking an alternative system to use in competing with the Morse companies.

The Morse interests relied on their patent monopoly in contesting the House and Bain patents. As the first telegraph inventor in the country, Morse sought to take advantage of the greater protection afforded by the passage of the 1836 patent act. Under the old system, in which applications were

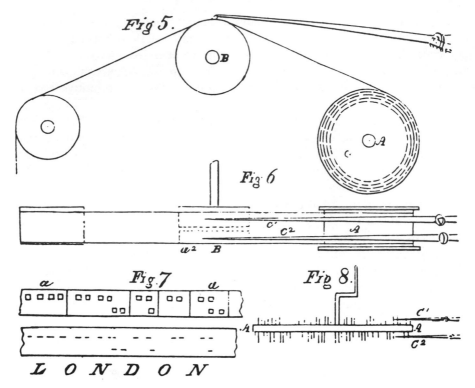

Figure 2.13. In the transmitter of Alexander Bain's automatic telegraph system, perforated paper passed over a metal drum, allowing a lever or roller to make contact with the drum and thus open and close the circuit. At the recording instrument, corresponding marks were made on electrochemically treated paper as the current passing through the recording lever caused it to discolor. (*Source:* Turnbull, *Electro-Magnetic Telegraph,* 38.)

merely registered and a patent automatically issued, inventors found it difficult to protect themselves against infringers as patents carried no presumption of validity. Under the new law all applications had to be examined for novelty and could not issue as patents unless they were considered new and useful. In this way, the temporary monopoly granted with a patent had much stronger standing before the courts when inventors or their assignees sued for infringement. Most important, Congress established a patent office and erected the necessary bureaucratic apparatus to administer and enforce the examination system.[56]

Amos Kendall and F.O.J. Smith sought control over telegraphy by claiming that the Morse patent controlled telegraphic transmission employing electromagnets. The unpopularity of monopolistic claims such as these in a society that mistrusted power and special privilege allowed O'Rielly to gain a great deal of public support in his attacks on the Morse lines. The ex-newspaperman turned to the newspapers where he published letters by himself and the various inventors challenging the Morse claims and gained the support of many newspaper editors. A practiced publicist, O'Rielly also published circulars that played on American themes of national pride as well as fears of monopoly.[57] O'Rielly, however, had no strategy for turning public opinion into a weapon that he could use against the Morse interests. He did not attempt to gain favorable legislation in Congress or in state legislatures and public opinion was a poor strategy to use in law courts, although he challenged the Morse patentees to bring suit against the inventions he supported.

Ultimately, the law courts proved more important than the court of public opinion in determining the fate of the various telegraph patents. Unlike O'Rielly, the Morse companies pursued a well-conceived strategy of price competition, construction of additional lines, rejection of interconnection, and, most important, patent infringement suits. In order to improve his already strong patent position, Morse and his partners had his original patent reissued in 1848 with a new claim covering broadly the art of telegraphing by electricity. The first suit, heard in 1850, was brought by F.O.J. Smith, whose New York–Boston line was being challenged by the House line established by Hugh Downing. The U.S. Circuit Court for the District of Massachusetts rejected Morse's broad claim from his reissued patent and upheld the House patent as a separate and distinct invention.[58]

The Bain automatic, which was being used on several lines, including O'Rielly's, was challenged next. Benjamin French, president of the Magnetic Telegraph Company operating between New York and Washington, brought suit against Henry Rogers, who represented a line between these same cities that employed Bain's system. Much of the evidence used in the House case was introduced again in this case heard in the U.S. Circuit Court for the Eastern District of Pennsylvania. This time, however, the court gave a much wider interpretation to the patent law. Basing its ruling on the constitutional provision for the promotion of the useful arts and congressional patent acts that placed the discovery or invention of "a new and useful art" first among those deserving protection, the court held that Morse was entitled to broad claims to control the art of telegraphy.[59] Critics considered the decision in the Bain case as an overly broad interpretation of patent law and some claimed that it was the result of an agreement to consolidate the Morse and Bain lines. The Bain cause may also have been hurt by the attempts of Henry Rogers, the former Morse operator who managed the Bain line between Washington and Baltimore, to improve the operation of the Bain system by adopting the Morse local circuit.[60]

The Supreme Court decision against O'Rielly in a suit on the Zook and

Barnes "Columbian" telegraph suggested that on appeal the Bain case would likely have been overturned. While the court held the invention to be an infringement on Morse's patent, it denied his broad claims to control the art. However, the costs of litigation and the uncertainty of the outcome caused a consolidation of the Bain and Morse lines following the original ruling of 1851. By the time the Supreme Court ruled in 1853, the Morse interests controlled the Bain lines. Only the House lines remained as a significant competitor of the Morse, principally on the Boston-Washington and New York–Buffalo lines.[61]

The Morse interests found that the technical advantages of their system were most important in competing with the House lines. Although the House printing telegraph was faster, its complex construction and transmission problems over long lines made it less reliable and desirable than the Morse apparatus, especially after the latter was greatly simplified after sounders replaced the registers. The Morse system of telegraphy, although undergoing significant change in the decades following its introduction, retained the core characteristics developed by Morse and Vail. This included a key, an electromagnetic relay, a register to record the coded message on paper tape, and the Morse code for translating electric signals into a verbal message. This code consisted of a system of dots and dashes that enabled operators to send messages by a simple sequence of opening and closing the electric circuit. Miles of uninsulated iron wire strung on ranks of wooden poles and attached to insulators linked terminal and way stations, while batteries powered the system. An operator sent a message by "opening" the key, causing a "break" in the circuit. This break was registered at the end of the circuit by the armature of the relay and register, signaling to the operator to await a message. The sending operator would then manipulate the key, thereby transmitting a sequence of dots and dashes that were recorded as indentations on the paper tape running through the register at the other end of the line. In the closed-circuit system used on American lines,[62] batteries located at the terminal stations provided the line current, while intermediate offices used batteries to operate their own local relays and registers.

By the mid-1850s sounders largely replaced the more complex Morse registers, although registers continued to be used into the 1870s, especially in small offices. Sounders were electromechanical instruments that operated on the same principle as the registers, but which received the message via the click made by an iron armature striking the core piece of an electromagnet when the sending key broke the circuit. Operators "read" the clicking armature and immediately transcribed the message rather than waiting for the register to record it on paper tape as dots and dashes for later transcription. The use of the sounder substantially increased transmission speeds and reduced errors. Furthermore, the simple mechanism of the sounder cost less

Figure 2.14. Sounders began to replace the Morse register after operators discovered that they could hear messages by listening to the clicking of the armature against the register's electromagnet. In these devices the volume of sound was increased by attaching to the armature a metal lever that struck against a metal sounding frame. (*Source:* Prescott, *Electricity and the Electric Telegraph,* 437.)

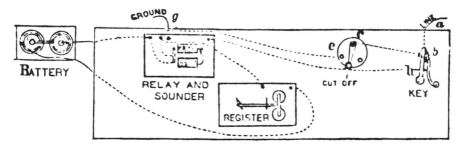

Figure 2.15. The basic Morse telegraph instruments commonly found on an operator's table were a key, a relay and sounder, and often a register and a circuit changer. The local battery would have been next to the table. (*Source: Harper's New Monthly Magazine* 47 [1873]: 349.)

to build and maintain than either the register or the House printing telegraph. By the end of the decade, the Morse system dominated American telegraphy.[63]

Though printing telegraphs had disadvantages when compared to the Morse, they did provide effective competition during the early 1850s in combination with other competitive strategies. The Western Union Telegraph

Company, which came to dominate the industry, began life as a House printing telegraph company and was able to use that system not only to compete with Morse but to gain access to his superior system. Established in 1851 as the New York and Mississippi Valley Printing Telegraph Company, the company rapidly expanded through advantageous agreements with several railroads. These railroads realized the value of telegraph communication for managing railroad traffic and, therefore, agreed to build telegraph lines, provide offices, transport telegraph equipment and employees, and deny access to the railroad rights-of-way to other telegraph companies. In return, the telegraph company issued stock to the railroads, maintained the lines, supplied operators to the offices, and gave fee priority service for railroad messages. The telegraph company, renamed Western Union in 1855, used railroad contracts to gain control of other major companies in the Midwest by the end of the decade. Through consolidation Western Union obtained licenses for the Morse patents, which proved more suitable for railroad operations than the House, and soon commanded the major western telegraph routes.[64]

While Western Union was consolidating the western telegraph lines, the American Telegraph Company, organized in 1855, secured enough eastern companies, along with crucial links to Nova Scotia and the planned Atlantic cable, to dominate the telegraph market in most of the Atlantic coast states. As part of their strategy, the incorporators of the American Company acquired the rights to David E. Hughes's improved printing telegraph from Daniel Craig, general agent of the Associated Press. The American Telegraph Company envisioned the Hughes instrument as a valuable weapon in competing with both Morse and House lines. The new printing telegraph appeared to have significant advantages over the House. Drawing upon his experience as a music teacher, Hughes used well-known acoustical laws to regulate the speed of all the instruments on a line. He designed a vibrating spring governor to control an escapement and regulate the speed of each instrument. By vibrating the governors of all machines on a line at the same frequency he achieved synchronous action. The Hughes telegraph was also much faster, as he designed the inked typewheel to press against the paper as it turned rather than stopping to print by pressing the paper against a ribbon as did the House. Hughes's design also required less battery power and employed simpler mechanism than the rival printing instrument.

The managers of the American Company soon discovered, contrary to Craig's sanguine pronouncements, that the Hughes system still contained bugs. Hughes had designed the printing instrument to work with two trains of gears, one controlling the typewheel and the other the printing mechanism. The lightweight construction required by its high-speed operation and the use of a weighted pendulum to drive the machine (rather than requiring

Figure 2.16. This photograph of David Hughes's printing telegraph shows its simpler mechanical design compared to House's instrument (see Figures 2.12 and 2.13). The transmitter required only a keyboard operator as weight-driven clockwork gearing replaced the hand-crank and other complicated mechanism of the House printer. Hughes's background as a music teacher is evident in the transmitting cylinder with pin projections, which resembled the design of a music box, and in the vibrating springs that resonated at a set frequency in order to synchronize transmitting and receiving instruments. (*Source:* Smithsonian Institution photo 46777.)

an operator to turn the instruments) caused the mechanism to break down under heavy use. To solve this problem, the American Company employed George M. Phelps, a Troy, New York, manufacturer, who devised an electro-motor governor that improved the synchronism and designed the gearing of the typewheel shaft and printing mechanism to be driven by the same motor and governor. Phelps also replaced Hughes's keyboard with one similar to that of House, thus giving the instrument its name—the "combination printer."[65] As in the case of Western Union, the new printing telegraph was to prove most useful in combination with other competitive strategies.

By the late 1850s, the need for interconnection to transmit long-distance messages between areas controlled by opposition companies, the desire for access to Morse system technology, and cutthroat competition combined to encourage consolidation in the telegraph industry. In addition, New York State incorporation laws gave decided advantages to the New York–based Western Union and American Telegraph companies in seeking control of

Figure 2.17. Mechanic George Phelps modified the Hughes printer by incorporating the compressed air printing system of the House printer, as well as a new electromagnetic governor of his own design, in what became known as the "combination printer." (*Source:* Prescott, *History, Theory, and Practice,* frontispiece.)

companies chartered in the Midwest. When western telegraph systems began joining networks growing out of Philadelphia, the New York legislature became concerned about the state's competitive position. In response it passed liberal incorporation and other laws to give state-chartered telegraph companies advantages in consolidating western lines and linking them to New York City, thus providing advantages for the city's businessmen and bankers.[66] The experience of telegraph entrepreneurs during the industry's first decade reinforced the tendency toward consolidation rather than cooperation. The failure of both the Morse patent interests and the O'Rielly lines to maintain cooperative alliances led subsequent system builders to expand their systems by combining independent lines into a single corporation through takeovers and consolidation.

Actions by the federal government combined with corporate goals to further encourage consolidation. In 1857 American Telegraph and Western Union joined with four other companies in a cartel agreement known as the "Treaty of Six Nations." This treaty carved the market into clearly defined territories and required interconnections exclusively for parties to the agreement. The agreement also divided the cost and rights to the new printing telegraph designed by David Hughes and subsequently improved by George

M. Phelps, cross-licensed the Morse and House patents to the six companies, and denied patent rights to companies not in the cartel.[67]

Like most cartels, the Treaty of Six Nations proved to be an unstable arrangement. It effectively ended in 1860 when Western Union broke ranks and won a $40,000 government subsidy to construct a transcontinental telegraph. The transcontinental telegraph had been urged on Congress for nearly a decade and in June 1860 a bill was finally passed. While the cartel initially supported the bill, it refused to bid on the line because of the low subsidy provided. Western Union therefore broke with the cartel in the belief, proved correct, that business on the line would be profitable. The government subsidy and rights-of-way provided significant advantages for Western Union in case its rivals attempted to build a competing line.[68]

With the completion of the government-supported transcontinental line in 1861, Western Union began moving toward a national system of telegraph service. The opening of the Civil War in that same year provided new government support that enabled the company to gain additional advantages over its rivals and to consolidate such a system following the war. While the cutting of Southern telegraph lines reduced the American Telegraph Company's network of lines by nearly half, Western Union, with its network of east-west lines in the North, experienced increased demand. In response, Western Union built new lines, equipped new offices, and increased staff and services. Large government subsidies supported much of this expansion. The company also contracted to build many of the lines wanted by the military, and its general manager, Anson Stager, became superintendent of military telegraphs for the Union side while continuing to run the company. Western Union emerged from the war with 44,000 miles of wire, more than the combined total of its two major competitors, the American Telegraph Company and the United States Telegraph Company, a new challenger in the field. The decline in war-related demand after the Union victory put great pressure on many telegraph companies and Western Union aggravated the financial difficulties through price wars. By 1866 Western Union absorbed both of its principal rivals. Although several regional companies remained independent and new companies almost immediately challenged the consolidated giant on its profitable routes, Western Union's size and national network enabled it to remain dominant.

The emergence of large interstate corporations in the telegraph industry created a new context for invention and presaged a general trend that became common by century's end. Undergirding this new context was a corporate culture being created by leaders in economic, educational, and government institutions who perceived a need for rational, "scientific," coordination of the economy and of the nation.[69] At the heart of this corporate culture were centralized management structures necessitated by the increasing "volume of economic activities . . . that made administrative coordination more efficient and more profitable than market coordination."[70] Large corporations began to rely less on the invisible hand of the market and more on what Alfred Chandler has called the "visible hand" of modern management. Though much of their rhetoric celebrated the virtues of a free market, those embracing centralized management were in fact turning away from classical economic liberalism.

Corporate consolidation and centralization were also spurred by the growth of large-scale technical systems. During the nineteenth century, the railroad and telegraph industries were the first to develop such systems as national markets emerged that transcended local and even regional systems. The need to coordinate and control far-flung operating units, which had emerged by mid-century in these industries, prompted corporate managers to initiate administrative reform in order to improve operating efficiency and achieve the economies of scale offered by their expanding technical systems. In the closing decades of the century, modern management practices were also initiated in manufacturing and marketing firms expanding to take advantage of perceived economies of scale created by new production and distribution technology. As professional managers dominated corporate decisionmaking, they adopted strategies forsaking short-term profits in order to ensure the long-term growth and stability of their firms.[71] In telegraphy the control of new technologies became an important part of such strategies and exerted powerful influences on the industry's technical community.

3

Invention and the
Telegraph Technical Community

THE INITIAL development of telegraph technology relied on innovators who had come to telegraphy from the outside: Morse was a painter; House was a dilettante with independent means and an interest in invention and electricity; and Bain was a clockmaker who turned to telegraphy after working on electric clocks. The diversity of their inventions reflected the variety of conceptual models supplied by their mixed backgrounds. Once these new systems became the basis of competing companies, they provided a conceptual framework that dominated the work of future telegraph inventors, who drew on their practical experience as telegraph contractors, managers, manufacturers, and operators in designing their inventions.

As those individuals who possessed practical experience in operating the nation's telegraph system came to play the central role in the industry's technical development, they created a technical community in the industry's operating rooms and manufacturing shops. This technical community developed in a pattern seen in many other American industries dominated by mechanical technology. A "shop culture" emerged in which practical experience, gained through the operation and manufacture of instruments, was supplemented by reading in technical literature. Shop culture embraced those elements of emulation that characterized mechanical technology and influenced invention in nineteenth-century America.[1]

Such patterns of emulation also stimulated the inventive efforts of members of the telegraph community. Few inventors actually worked alone and none worked in complete isolation. Conceptualization was often stimulated by access to new information. And the construction, testing, and redesigning of apparatus necessary for practical application almost invariably required an inventor to seek the assistance of others, whose own contributions often altered the original design. The power of the popular heroic model of invention, which treats initial conceptualization as the true inventive act, has caused the more methodical activities involved in the development of a practical invention to often be treated as ancillary, thus obscuring the diversity of other contributions often necessary to the successful inventive enterprise.

Morse found it necessary to draw on the talents and knowledge of others by bringing together pre-existing scientific and technical communities. In the process he began the creation of a specialized technical community. As the telegraph industry grew, this specialized technical community would provide the necessary resources of knowledge and skill for telegraph inventors. The knowledge of electrical science that Morse gained through personal contacts became available to later inventors in more formal fashion in the industry's technical publications, while the talents of a skilled mechanic such as Vail would be found readily at hand in the telegraph manufacturing shops.

Although telegraph invention usually required some knowledge of electricity, practical experience and knowledge of mechanical movements often proved more valuable than advanced knowledge of the science of electricity. Those operators responsible for maintaining lines and instruments could acquire a good practical knowledge of electricity as well as of the mechanical operation of their equipment. The operating room served as a technical school in which the would-be inventor could secure necessary knowledge of telegraph technology. The industry's machinists gained their own electrical knowledge through practical experience in building and testing telegraph equipment and in working with inventors. When the problem was essentially mechanical even mechanics unschooled in electrical science sometimes produced important improvements. Because telegraph design remained largely mechanical in nature throughout the nineteenth century, the machine shop was an important inventive center in telegraphy. One important result of this was that the telegraph technical community acquired a specific geographical focus as it became centered in those major eastern and midwestern cities that boasted telegraph manufacturing shops.

While the machine operators, manufacturers, mechanics, engineers, and inventors who made up the telegraph technical community all possessed some degree of technical knowledge, only a small percentage sought to study and improve technology. Those who did believed that their increased knowledge and proficiency in the technical operations of the system might bring them personal advancement. Working within the structure of the telegraph industry they usually focused on technical problems plaguing existing systems. Unlike the industry's pioneers, most saw invention as a way to move into management positions as chief operators, office managers, superintendents, and, with the expansion of technical staffs in the 1870s, engineers.

The Emerging Technical Community

The rapid growth of the early telegraph lines produced a need for minor inventions that would improve their working efficiency. Contractors responsible for building these lines were responsible for most of the early inventions concerned with line efficiency. They primarily concentrated their efforts on improving the wires and their insulation. Devising effective insulators proved to be the most difficult technical problem facing the new industry. Initially, the problems of insulation were little understood; F.O.J. Smith even denied the need for insulation, with disastrous affects. Henry O'Rielly, on the other hand, ordered the wire itself coated with tar as insulation, while Amos Kendall ordered every other pole cut down in order to reduce the number of points at which insulators were needed. Within a short time, however, other inventors began to devise insulators on more sound principles. Insulators needed to be made of highly insulating materials, resistant to moisture, and not easily cracked or affected by exposure to heat and weather. They also needed to provide the greatest possible insulating distance from the pole while retaining sufficient strength to support the wire. All of the materials available to inventors were susceptible to one or more of these difficulties and inventors attempted to limit their effects on the insulation of the line. Ezra Cornell's glass drawerknobs proved to be a prototype as glass insulators became popular in America, while Europeans favored earthenware. Although glass became common in the United States, wood, earthenware, porcelain, white flint, hard rubber, and iron all had their advocates. Inventors also modified the form of the insulators and sought to protect them against moisture, heat, and breakage while retaining their insulating properties.[2]

Contractors also sought to improve the wire conductors as a means of increasing transmission efficiency. Again limited knowledge of conducting materials plagued the early inventors. Copper, though a good conductor, was not strong enough. Most lines used uncoated iron wire which, though susceptible to rust, was much cheaper and stronger. Marshall Lefferts, who became interested in telegraphy as a supplier of wire, introduced European galvanized iron wire on the Bain lines; this proved to be the best conductor for strength, durability, and line efficiency. It improved the working of lines using the Bain instrument and, following their takeover by the Morse companies, also proved best for the Morse lines. However, probably owing to their greater cost, galvanized wires were not as widely used as uncoated iron.[3]

Unlike the first contractors, many of the early operators were attracted to careers in telegraphy because they already had prior knowledge of the science of electricity gained from reading and private study, particularly as literacy in the English language was an important requirement for telegraph

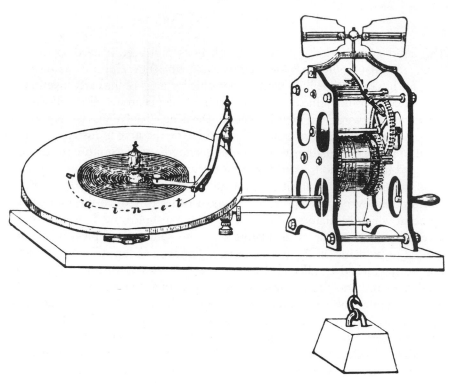

Figure 3.1. Telegraph operators Henry Rogers and Charrick Westbrook modified the Bain automatic by replacing the strips of recording paper, which were saturated with chemicals and thus easily torn, with a circular disk upon which they placed a single sheet of chemical paper. This modification became standard on Bain automatic telegraphs used in the United States. (*Source:* Prescott, *History, Theory, and Practice,* 129.)

operators. Henry Rogers, who was the first operator hired to work the Morse system, was typical of the inventors who emerged from the early operating corps. Educated at St. Mary's College in Baltimore, Rogers became interested in the new electric telegraph and wrote Morse asking for a job. He was appointed the Baltimore operator of the experimental line to Washington. Rogers continued serving in that capacity when the line was briefly run by the Post Office department and during the early years of its management by Morse's Magnetic Telegraph Company. Though an original stockholder in Magnetic Telegraph, Rogers went on to became superintendent of a new company operating the Bain automatic system. Under Rogers's direction, important changes were made to the Bain instruments. Unfortunately, existing records do not indicate the extent to which Rogers himself was responsible for changes in the transmitter and receiver of the Bain system. But the

Figure 3.2. The telegraph operating room provided a practical school of electricity. (*Source:* Shaffner, *Telegraph Manual,* 461.)

perforated transmitter was replaced with a finger key, similar to the one used by Morse, and the strip of chemically saturated recording paper was replaced with a circular sheet of paper on a horizontal metal disc. Rogers was directly responsible for the adaptation of the Morse local circuit to the Bain lines, thereby improving transmission to intermediate stations. With Charrick Westbrook, a Bain operator, Rogers invented a new automatic electrochemical telegraph in which the paper recording surface was entirely replaced by a brass disk, and later he worked with manufacturer Samuel Bishop on wire insulation.[4]

The most prominent telegraph inventor of the mid-nineteenth century, Moses Farmer, also came to telegraphy with formal knowledge of the science of electricity. Three years after graduating from Dartmouth College in 1844, Farmer presented a series of lectures on electromagnetism during which he also demonstrated a model electric railroad of his own design. These lectures brought him to the notice of F.O.J. Smith, who was building lines for the New York and Boston Magnetic Telegraph Company. During the next several years, Farmer superintended and repaired lines between Boston and other cities in Massachusetts and also conducted a variety of electrical experiments. By 1851, he became associated with William F. Channing, son of the noted clergyman William Ellery Channing, in the development of the first fire-alarm telegraph system and made important contributions to the mechanical as well as electrical design of the system. Over the next two decades Farmer patented a series of telegraph and other electrical inventions that enabled him to become the country's first independent electrical inventor.[5]

As the industry grew and the operating and manufacturing corps expanded, those engaged in telegraphy had less initial familiarity with electricity, but educated themselves through day-to-day experience in its practical use. They also read about electrical science as well as practical telegraphy in the pages of a new, specialized technical press. This growing technical literature and a sense of pride in their specialized knowledge inspired many members of the emergent technical community to undertake further study. The community also encouraged experiments to find solutions to problems plaguing the new technology. In an era when electrical engineering had not yet emerged as a profession, they made no distinction between invention and engineering in their attempts to improve the technical operation of telegraph lines.

For most inventors the telegraph operating room played the most important role in their electrical education. It was through their daily duties that most operators gained experience with both the electrical working and mechanical construction of the apparatus. They frequently replaced wiring in defective instruments, and some operators were capable of repairing the instruments as well. Operators often modified sounders in order to vary the sound to suit their personal preference, and the lack of central engineering departments and of uniform engineering practice also meant that they often connected apparatus in different ways. Operators became familiar with the changing electrical qualities of the line as they learned to adjust their instruments to compensate for faults in the line caused by poor insulation or bad weather. They also maintained the batteries and inspected the lines for faults. In the process they gained a practical knowledge of electricity that aided them in their inventive work.[6]

David Brooks was typical of the early telegraph employees whose electrical knowledge grew directly out of experience. Brooks first entered the telegraph industry in 1845 when he worked on the construction of one of the earliest lines and assisted O'Rielly with his unsuccessful attempts at insulation. Following the Civil War, Brooks invented one of the most widely used insulators. As a telegraph office manager and line superintendent he earlier focused his inventive energies on other problems. In the areas of the United States where the first lines were built, thunderstorms were a common occurrence and lightning wreaked havoc with the operation of the lines. Brooks sought to solve this problem about 1847 by wrapping paper around the gas pipe in the telegraph office and then connecting the pipe to the main line with a piece of copper wire. In this way he discharged the lightning through an earth connection and protected the instruments. Though Brooks never reduced this plan to a patentable device, it embraced the basic principles found in the lightning arrester designs patented later.[7]

Charles Smith, who designed the lightning arrester most commonly used in American telegraph offices, was another practical electrician whose knowledge came from his experience as an operator. Smith first became associated with telegraphy when he assisted Morse with some electromagnet experiments in Washington, D.C., in 1845. He subsequently became a Morse operator and, like many others, learned about the effects of lightning from direct experience.[8]

Operators such as Brooks and Smith gained reputations for their knowledge of telegraph technology that was further enhanced by their inventive efforts. By contrast, many of those who entered the industry as line repairers or constructors exhibited a woeful lack of electrical knowledge that caused frequent complaint.[9] The superior knowledge of electricity possessed by operators was a direct consequence of the nature of their work. Transmitting messages over a telegraph line enabled them to develop a practical familiarity with electricity. It also required a high degree of literacy, which encouraged many operators to improve their skills through reading books and journals containing information about electricity and telegraphy.

By the 1850s this literate workforce encouraged the emergence of a specialized literature. The first journal catering primarily to the telegraph industry was the *American Telegraph Magazine* published by Henry O'Rielly in 1852. Though O'Rielly began the journal primarily as a weapon in his dispute with the Morse interests over the extent of their patent monopoly, he and editor Donald Mann presented much information about the telegraph industry both in the United States and abroad. Perhaps most significant from the standpoint of technical development, they recognized that the progress of telegraphy, which they believed "indispensable to the welfare of society," depended on the knowledge and skill of telegraphers. Toward this end they also sought to publish materials of interest to telegraph employees including "all valuable information, illustrative of the Origin and Progress of Electric and Chemical Science in their connection with Telegraphic Art—reproducing many articles by eminent authors, which are now accessible only through tedious and expensive research."[10] They also published articles concerning new inventions and encouraged continued improvement in the art, as well as calling on telegraphers to furnish "notes of their experience, in cases where that experience corrects [electrical] theories commonly entertained."[11] The *American Telegraph Magazine* and its two successors, James Reid's *National Telegraph Review* and Taliaferro Shaffner's *Telegrapher's Companion,* were all short-lived; none lasted more than a year. Nonetheless, these were important sources of technical and business information about the telegraph industry.

They were also early attempts to forge a national telegraph community and served as models for the national telegraph journals published after the Civil War.[12]

In the late 1850s, prominent members of the growing technical community began to publish manuals detailing the history and operation of the various telegraph systems, and providing basic information about electricity. The first of these manuals appeared in 1859 and was written by Taliaferro Shaffner, former editor of the *Telegrapher's Companion*. Shaffner, who constructed and managed a number of early lines, included details of the administration as well as the operation of European and American telegraphs. He also described new inventions, so as to encourage further improvements, and provided a list of European books on electricity and telegraphy to encourage further study by operators.

Although Shaffner encouraged operators to acquire knowledge, and recognized that such knowledge would enable them to operate the telegraph systems "nearer to a state of perfection," he did not believe that such information was indispensable to a successful career. On this issue Shaffner's views were shaped by the democratic values of antebellum America. While European telegraphs required special training of their operators, "in America, there are no such necessities existing. Labor, in whatever branch, cannot be superior to that of another. This equalization is a fundamental and cardinal virtue in American institutions." He thus concluded that the average American operator had "no superior," even though most had little training beyond a basic grade school education "and a few months' practical service as a manipulator." Shaffner's views were probably influenced as well by cost considerations. He considered the cost of specialized training such as that received in Europe "too great for the requirements of the service," unless it could be achieved at "moderate expense."[13] This typically American analysis placed responsibility for both self-improvement and the improvement of technical operations on the individual operator.

The other important early telegraph manual, George Prescott's *History, Theory, and Practice of the Electric Telegraph*, appeared in 1860. Unlike Shaffner, Prescott considered the telegraph to be a "true science" that required operators to be more than skilled manipulators of their instruments. He argued that each operator should "possess a knowledge of the technical part of his service, to foresee natural phenomena which can influence transmission, and understand the derangements which take place so frequently upon posts, wires, and other apparatus of the line, determine their causes, repair accidents, in the majority of cases, and furnish, when there is need, a fund of general knowledge upon the subject, to meet all emergencies."

It was thus necessary that "he be initiated into the laws and properties of electricity, that he may render himself entirely competent to comprehend

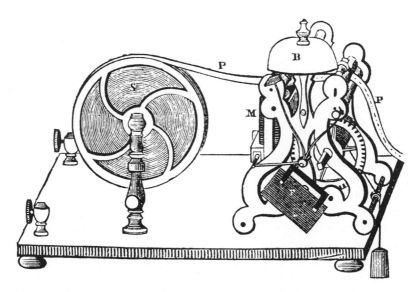

Figure 3.3. Telegraph manufacturers such as Daniel Davis sometimes produced manuals of electricity and telegraphy in which they advertised their own instruments, such as this version of a telegraph register with an alarm bell to alert the operator to an incoming message. (*Source:* Davis, *Manual of Magnetism,* 203.)

all the laws respecting the transmission of electric currents, and that he know perfectly all the details of construction of the batteries, instruments, &c."[14] But Prescott did not go so far as to believe that this "initiation" should be conducted by telegraph companies. Instead, he cautioned them to be careful in their selection of employees; acquisition of requisite knowledge still depended on the personal ambitions of individual operators and it was toward that end that Prescott presented his work.

The hierarchy of skill and status within the operating room helped to sustain ambitious operators and encourage invention. At the top of the hierarchy were office managers, who reported to divisional superintendents and their staffs, followed by chief operators, circuit managers, traffic chiefs, and the operating and clerical staffs. The technical expertise needed by chief operators and circuit managers, who maintained the circuits and were responsible for line breaks that interrupted service, provided them greater prestige and mobility than traffic managers who controlled the flow of messages. Operators themselves were clearly above clerks and messengers, full-time operators above part-timers, main-line operators above branch or local circuit operators, fast senders and receivers above those who worked more slowly, and press-wire operators above all the rest. In this context, experience counted for far more than any formal training. Most operators developed their skill through a combination of techniques, using practice keys and handbooks for self-instruction and serving informal apprenticeships with established operators.[15] However, while all possessed some degree of technical knowledge, only a small percentage of operators sought to study and improve telegraph equipment. This latter group included ambitious operators who increased their knowledge and proficiency in the technical operations of the system in order to gain personal advancement. Electromechanical invention could serve as a source of career mobility because it allowed operators to demonstrate both the technical expertise and the personal initiative necessary for such management positions as chief operators, office managers, and superintendents.

While the operating room provided a school for electricians and provided a common experience that helped to forge an operator culture, the central role played by the machine shop in telegraph invention, as in many other American industries, led many inventors to leave operating for manufacturing. Charles Chester's career illustrates the growing importance of the manufacturing shop for telegraph invention. Chester graduated from Yale in 1845, briefly studied medicine, and then became a telegrapher after the failure of his father's business. Chester may have been attracted to the new industry because of the knowledge of electricity he acquired at Yale. By 1852, his "scientific attainments, and mechanical knowledge" had become well-known and he was appointed superintendent of the telegraph manufactory recently established by John Norton in New York City. Chester, who with his brother John took over Norton's shop a few years later, commenced a long career as a telegraph manufacturer-inventor and produced a number of important inventions. The brothers Chester became noted for the assistance they provided inventors.[16]

Chester joined several other manufacturers who began work in the 1850s to meet the growing needs of the telegraph industry. Most of these first telegraph manufacturers came to the industry from other pursuits—primarily instrument- and clockmaking. One of the original manufacturers of instruments for Morse was Daniel Davis, a Boston philosophical instrumentmaker who already had extensive interest in electrical science and technology. During the 1830s Davis constructed experimental apparatus for Charles Grafton Page, one of the leading American scientists working on electricity. Davis later manufactured several of Page's electrical devices as laboratory apparatus. Davis was also a pioneer American daguerreotypist and devised a new technique for electroplating the silver surface on daguerreotype photographic plates in the early 1840s. His growing experience with electrical science and technology led Davis to publish the first American manual of electricity in 1842, and in 1844 he manufactured some of the first instruments for Morse.[17]

Davis was one of a large number of mathematical, optical, and philosophical instrumentmakers who made a living in Boston, which also became an early center of telegraph invention. Among those who manufactured electrical devices in the city during the 1850s were instrumentmakers J. M. Wightman, N. B. Chamberlain, and E. S. Ritchie, all of whom probably manufactured telegraph instruments as well. Davis's shop was later run by Thomas Hall, who was still an important manufacturer in the early 1870s. Another Boston firm from the 1850s, Hinds and Williams, was later succeeded by Charles Williams, Jr., who became one of the most prominent of the telegraph manufacturers. For a time in 1858, Moses Farmer also manufactured telegraph equipment in Boston.[18]

Boston remained the major telegraph manufacturing center in the country until after the war, when New York and Philadelphia also became important centers. However, some important manufacturers did appear in those cities before the war. William Clark was trained in England as a mathematical instrumentmaker before coming to Philadelphia where he took up clockmaking. Clark and his son John manufactured some of the instruments used on the first line between Washington and Baltimore and continued to manufacture telegraph equipment. John Clark also became an important telegraph inventor, patenting a number of improvements in repeaters, relays, and other instruments.[19]

New York was the last of the three great eastern cities to acquire a telegraph manufactory when John Norton opened his shop in 1851. The Chester brothers took over Norton's shop about 1854, and George Phelps, another important inventor-manufacturer, moved to the city two years later. Phelps, who had made major improvements in the Hughes printing telegraph for the American Telegraph Company, became that company's principal manufacturer and moved his shop from Troy, New York. Phelps continued to serve as superintendent of the shop after Western Union took over American Telegraph following the Civil War and he made further important inventions.[20]

American Telegraph Magazine editor Daniel Mann's extensive description of John Norton's New York shop indicated the role that manufacturers could play in inventive activity. Mann noted that a line running to the company's store provided the means for "the *practical* testing of every register and magnet sent out." But it also allowed superintendent Charles Chester to test a new battery design under working conditions. Practical tests of new inventions were a common function of such lines. In the rear of the factory Mann found "a hospital and museum for veteran, diseased, and defunct telegraph instruments; and every variety of form, size, and principle, had there its representative." Design improvements that Mann "noticed in the finished machines" made by the shop may well have been influenced by the

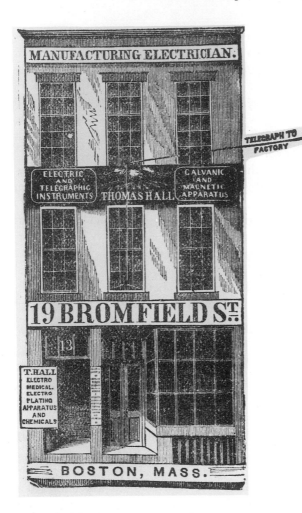

Figure 3.4. Besides test lines inside the factory, telegraph manufacturers often had lines between their shops and factories that could be used for experimental purposes. (*Source:* T. Hall, *Catalogue,* 2.)

study of older instruments such a "museum" allowed. Mann concluded his article by noting that it was Norton's intention, assisted by Chester, to not only "supply the demand for telegraph equipment," but "to bring forward several improvements and novelties, now in embryo, and to improve the unequaled facilities he has of doing everything for the telegraphic community that can be demanded."[21] Indeed, in his advertisement in the magazine Norton stated

> I am prepared to carry out any original design, alteration or improve-ment in the construction of instruments, &c. which may be suggested by the taste of the purchaser.
>
> Every person connected with the Telegraph is desirous of seeing it improved, and every inducement thereto will be held out by me, to the ingenious operator and mechanic.[22]

Individuals connected with telegraph manufactories were responsible for several improvements in the design of system components, such as batteries, keys, registers, relays, and switches. Charles Chester and Thomas Hall, for instance, both devised improvements in battery and relay design that came into common use, while the Philadelphia manufacturer John Clark worked on Morse registers, relays, and repeaters. Such improvements probably reflected competitive pressures felt by manufacturers as they sought contracts to supply equipment to the large intercity telegraph companies. They also represented a response to new technical problems created by industrial consolidation.[23]

The decade of consolidation leading to Western Union's national network presented a host of new technical problems related to transmission efficiency. The effective transmission distance of a telegraph signal presented one key problem for inventors. Because the strength of a signal was inversely proportional to the length of the line for any given strength of battery, practical transmission distances were limited to about 250 miles. When the telegraph network consisted of small companies, this was a minor problem as most through traffic was retransmitted by hand on separate lines. As these lines were brought together under a single firm, telegraph managers sought means of transmitting their messages directly over a continuous line. Inventors therefore set about devising means to retransmit messages automatically. In 1848, Charles Bulkley, superintendent of the first telegraph line between Washington and New Orleans, devised the first successful automatic repeater in order to send messages directly between the two cities. His repeater, which transferred the weak incoming signal to a new transmitting circuit with its own battery and then forwarded it to the next station, served as a basic model for the subsequent refinements made by other inventors, all of whom were telegraph operators, managers, and manufacturers.[24]

Repeaters became a general subject of interest during and immediately after the Civil War. Daniel B. Grandy, who became an operator at the close of the war later recalled that "one of the first efforts of the embryo 'electrician' in those days, used to be the invention of a new 'repeater.'"[25] The way in which existing repeaters served as a model for young inventors is suggested by Thomas Edison's experience. While working as an operator in Cincinnati in 1867, he filled several pages of a small notebook with drawings of the many patented repeater designs. Some of the drawings show slight alterations and some were of Edison's own designs. One of these was a repeater design for the Phelps combination printer that Edison published in the *Telegrapher*; another was a button repeater in which the direction was manually switched. The button repeater gained public notice in Franklin Pope's *Modern Practice of the Electric Telegraph* as a design that was "found to work well in practice" and which would "be found very convenient in cases where it is required to

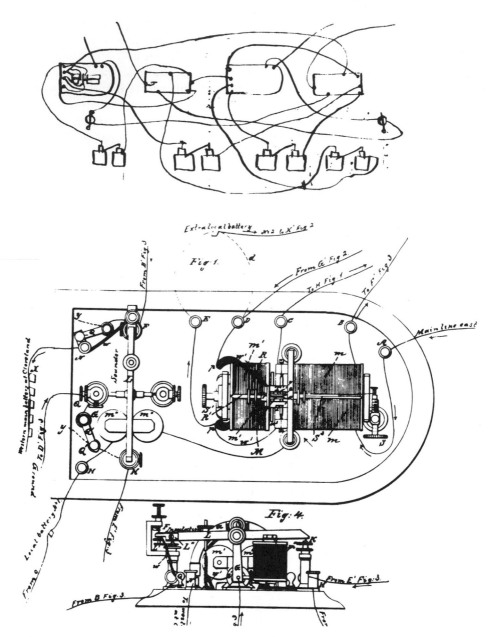

Figures 3.5 and 3.6. Thomas Edison drew this diagram of the office connections for George Hicks's self-adjusting repeater (seen in the patent drawing, U.S. Pat. 34,575) in 1867 while working as an operator in Cincinnati. Edison drew the sounders as separate pieces of equipment (the Hicks patent shows them on the same board), and also provided wires and keys to allow the repeater station to become a terminal for each circuit. (*Source:* PN-69-08-08, TAEM 6:772.)

set up a repeater in an emergency, with the ordinary instruments used in every office."[26]

Ambitious inventors also found that longer distances required related improvements in the design of relays, which switched the incoming signal from the main line to the local circuit, employing either a register or a sounder worked by a battery. Changing line conditions, created by atmospheric conditions and poor insulation, as well as long distances, caused the current to vary and required operators to adjust the springs on the relay armatures frequently in order to receive messages clearly. Inventors sought to develop self-adjusting relays that automatically compensated for varying line conditions. For example, Edison's 1867 notebook included several relay drawings, and the following year he published descriptions of his experiments and design for a self-adjusting relay in the *Telegrapher*.[27]

Inventors also designed improvements in circuit arrangements designed to economize on batteries as well as to improve transmission efficiency. Anson Stager, while an operator at Cincinnati in the 1850s, devised a plan for working several lines from the same battery. Stager's plan, which helped him achieve promotion, not only economized on batteries, but also allowed one operator to simultaneously transmit press copy or other messages to all the stations on the main circuits.[28] The completion of the first transcontinental line in 1861 and the consolidation and expansion of the wartime telegraph network by Western Union made further improvements in insulators and conductors essential as well.

The telegraph technical community that emerged in the 1850s and 1860s to respond to the industry's technical problems was forged by the conjunction of telegraphy's two shop cultures—those of the "ingenious operator and mechanic." Though most directly influenced by developments within the telegraph industry, particularly the emergence of Western Union as a national corporation, this community was also influenced by its interaction with larger economic and cultural patterns. Because of the nature of their work, handling messages from distant places, and their employment in one of the first large-scale national industries, telegraph operators were among the first groups to find their interests transcending the local community. This was part of a general pattern of American industrial development that disrupted many traditional, geographically based community structures, causing them to begin forging new communities less tied to a specific place.[29]

Even as they formed a national community based on functional relationships created by the possession of specialized skills gained through operating a new technological system, operators retained relationships of mutuality and sentiment similar to those found in traditional communities. Members of the telegraph "fraternity" kept in touch with each other by wire and through the pages of telegraph journals, and friendships developed among operators who never saw one another but who recognized each other's "touch" on the telegraph key. They also created a unique, extended subculture with its own lore, language, and attitudes toward the uninitiated.[30] The industry's manufacturing shops were affected by these same forces. Relations within the machine-shop community retained something of the personal obligations of the older craft community, at the same time those relationships became closely tied to an increasingly national marketplace. In the postwar era the telegraph community would find this marketplace dominated by a new phenomenon—the national corporation represented by Western Union.

Ideology and Education in the Corporate Context

Telegraphy initially proved a rewarding career for those young men who began as messengers and operators in the 1850s and 1860s. Entering a new industry, these men found that the diligent, hard-working, intelligent operator could advance into management positions. By the end of the Civil War, however, declining traffic and the consolidation of most important lines into Western Union caused opportunities and pay to decline. Even as telegraph business began to grow at the end of the decade operators found their circumstances little changed in the era of Western Union dominance. To meet the challenge of new competitors the company periodically engaged in tariff reductions accompanied by reduced wages. Furthermore, labor relations remained largely a local responsibility with operators dependent on the often arbitrary decisions of local managers regarding hiring, firing, work rules, promotions, and pay.[31] After a short-lived strike against Western Union in 1870 it took operators over a decade to reorganize against the industry giant.[32]

Professionally conscious members of the telegraph fraternity, led by James Ashley, the editor of the *Telegrapher,* responded to labor strife by focusing on self-improvement as the best means for both personal advancement and increasing the status of operators. In much the same way that craft artisans had been counseled earlier in the century to study and apply technical and scientific knowledge, Ashley and others argued that the acquisition of technical expertise and its application to invention was crucial to the career aspirations of operators. Through self-improvement individual operators would advance through the ranks, while continued innovation would promote competition and open up new opportunities.

Because technical improvements in telegraphy continued to be the product of practical knowledge, they were thought to be the province of any ingenious mechanic at the workbench or operator at the telegraph key whose experience and ambition provided the necessary knowledge and incentive to improve practice. Developments in America contrasted with those in Europe,

Figure 3.7. The masthead of the *Telegrapher,* the most important telegraph journal of the post–Civil War decade, was drawn by Franklin Pope, a leading telegraph inventor and engineer, who served briefly as its editor. (*Source: Telegrapher* 1 [1865]: 89.)

where state control of telegraphy and existing traditions of engineering professionalism fostered the development of elite corps of school-trained telegraph engineers.[33] For Americans, devising technical improvements in the form of new inventions was a means of getting ahead by becoming managers, manufacturers, or entrepreneurs; few were concerned with improved engineering practice or advancing theoretical knowledge.[34] Instead they expressed the democratic ethos of the American shop tradition fostered by mechanics' journals such as *Scientific American.*[35] In telegraphy, journals such as the *Telegrapher* also served as promoters of self-education and invention as the means to self-improvement and career mobility.

Operators were still expected to provide themselves with more sophisticated electrical knowledge.[36] However, few operators appear to have exhibited the private initiative necessary for achieving such knowledge. In a report on their visit to the United States in 1877, British Post Office Telegraph officials William Preece and Henry Fischer concluded that

> One or two operators are found in the larger cities students of Electricity and experimental investigations, but the great majority of Managers, Chief Operators, and Operators are simply contented with the *technical* knowledge their business demands. No steps are taken to increase their knowledge, and those who desire to advance higher have to do so at their own expense and with their own means.[37]

Most operators continued to rely on the primary electrical education they received in the telegraph office through performing their daily duties.[38] Indeed, Preece and Fischer noted during their 1877 visit that electrical responsibilities of office managers and operators was one of the "most marked features" of the American telegraph system. They found that managers and operators were responsible for the maintenance of the apparatus and batteries, and that duties performed by line inspectors in Britain were also largely

Figure 3.8. Operators were responsible for a range of technical responsibilities that included taking care of batteries. (*Source: Harper's New Monthly Magazine* 47 [1873]: 341.)

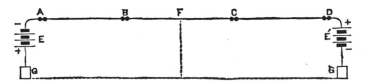

Figure 3.9. Operator responsibilities also included maintaining the electrical continuity of the line, and manuals such as Franklin Pope's widely read *Modern Practice of the Telegraph* included instructions for determining the location of line faults. (*Source:* F. L. Pope, *Modern Practice,* 75.)

discharged by the operating staff in the United States.[39] These technical responsibilities provided operators with their basic understanding of the electrical action of the conductors, insulators, batteries, and electromagnets used in telegraphy.

Telegraph machinists also relied primarily on practical experience and initiative in their electrical education. Some of their experience was gained in normal manufacturing work. Such activities as winding electromagnets, connecting wires in instruments, or testing apparatus on test lines in the shop provided practical electrical knowledge. Most shops undertook experimental work and some mechanics acquired more extensive electrical knowledge through working with inventors. Paul Seiler, who was foreman of the Electrical Construction and Maintenance Company in San Francisco, felt that his practical experience designing and building telegraph machinery allowed him to tell "by the drawing most of the time, without putting it into a practical machine," if a particular design would work electrically.[40] Thomas Watson, who assisted Alexander Graham Bell in his experiments on multiple telegraphy and telephony and later gained fame on his own as a telephone inventor and engineer, was another mechanic who received his practical electrical knowledge while working as an experimental machinist.[41] Charles D. Haskins, who left operating to find employment as a telegraph machinist, noted that while employed in the workshop of Welch & Anders he "acquired considerable information relative to electrical apparatus and the science of physics in general."[42]

These machinists were among many who also pursued a more sophisticated knowledge of electricity through private study and reading. Charles Batchelor, who assisted Thomas Edison in his telegraph and other experiments, was another. Batchelor was a British-born mechanic who began working at one of Thomas Edison's first telegraph manufacturing shops in Newark, New Jersey, in 1870. Until his employment with Edison, Batchelor had no experience with electrical technology. He soon began supplementing this with practical experimentation and reading and by 1873 he had become Edison's principal experimental assistant. Batchelor, whose work proved crucial to the success of many of Edison's telegraph, telephone, and electric lighting inventions, also patented his own electrical inventions.[43]

Telegraph machinists, as well as ambitious operators, could turn to a growing literature published in journals serving the operator community and in books written by prominent members of the technical community. The standard telegraph manuals of the day provided a systematic introduction to basic electrical science, as well as telegraph technology. Two of the leading American telegraph electricians, George Prescott and Franklin Pope, brought out important works republished in several editions during the 1870s and 1880s. Pope's *Modern Practice of the Electric Telegraph* was designed to give the operator a basic understanding of electricity and magnetism plus detailed knowledge of the working of the Morse system of telegraphy. For more detailed treatment of electrical theory and of important telegraph inventions, operators could turn to Prescott's *Electricity and the Electric Telegraph*. Prescott provided extensive information about French and German telegraph inventions. By using drawings and descriptions drawn from the technical literature of those countries, he made this information readily available to Americans.

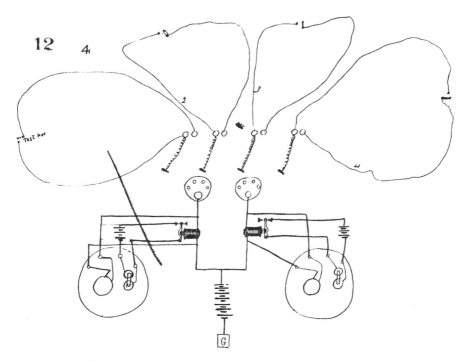

Figure 3.10. Charles Batchelor, a machinist who was Thomas Edison's principal experimental assistant, became a knowledgeable electrician who contributed his own inventive ideas, including this system for working four circuits with two recorders on the lines of Edison's Domestic Telegraph Company. (*Source: TAEM* 90:663.)

British works and shorter treatises on electricity published by the leading telegraph manufacturing concerns were also readily available. In his own manual, John Abernethy recommended several British books on electrical science, although he also encouraged "personal investigation and experiment" to supplement such reading. Many of the standard telegraph works, along with more purely scientific works, could be found in the free libraries of leading American cities. Cincinnati's Free Library was particularly noted for its fine collection of electrical works. It was here in 1867 that Edison discovered Michael Faraday's *Researches in Electricity.* Another Cincinnati operator, Charles Selden, who began working in the city in 1868 and later became an important telegraph inventor and electrician, also availed himself of the resources of this library.[44]

Less systematic knowledge could be gained in the pages of the leading telegraph journals. In addition to articles on electricity and magnetism, they provided a forum for the exchange of information about the state of the art, including accounts of new inventions and experiments conducted by readers.

The editors of the telegraph journals also spoke to the aspirations of inventors and, through their editorials, promoted the study of electricity and the development of new inventions.

The professionally conscious members of the technical community, led by the editors of the telegraph press, believed that an important means of improving the status of the profession was through the cultivation of new standards of technical proficiency founded on the study of the "science" of telegraphy. Recognizing that books on electricity were generally too expensive for most operators, an early telegrapher organization, the National Telegraph Union, had started the *Telegrapher* in 1864 as a means for the widely scattered and largely itinerant "fraternity" to communicate on personal or technical matters and as a place where the operator could turn for knowledge about electrical science.[45] Although the union foundered, the *Telegrapher* continued to publish until it combined with Western Union's *Journal of the Telegraph* in 1877. Under the editorship of Franklin Pope and James Ashley, the *Telegrapher* persisted in emphasizing to the telegraph community the importance of technical and scientific knowledge as a means of personal advancement and the improvement of the telegraphic art.

Western Union officers also encouraged operators to acquire technical proficiency. Western Union president William Orton founded the *Journal of the Telegraph* in 1867 as a means of communicating company policy and practice to the employees of Western Union's widely flung system. He also saw it as a means "to impart the necessary [scientific] education to enable those who are sufficiently interested in their future advancement in life, to fit themselves for the positions which will be open to them when qualified to fill them."[46] Like the *Telegrapher,* Western Union's journal, under the editorship of long-time telegrapher and former editor of the *National Telegraph Review* James Reid, presented articles on telegraph technology and new inventions. It also encouraged letters regarding technical questions in the belief that "they lead to experiment" and provided important information from the ranks for those in authority.[47] When the *Telegrapher* was combined into the *Journal,* James Ashley continued as editor to encourage those "who engage in telegraphic service to seek to make themselves proficient in the telegraphic art, and in the science upon which it is based." Ashley advised such study in the belief that "the development of the telegraph art . . . devolves upon the educated and skilled electricians and telegraphers, and future honors and rewards to be obtained in this direction must accrue mainly to them."[48]

Other telegraph journals followed the example set by these early postwar periodicals. The *Operator,* which was begun in 1874 primarily as a fraternal paper containing news about operators and entertaining articles, began including technical and scientific articles the following year, most of which were prepared by its science editor, Thomas Edison. Later contributions to this journal were made by noted electrician Thomas D. Lockwood, whose notes and queries column was subsequently published as *Electricity, Magnetism, and Electric Telegraphy.* By 1884, when publisher William Johnston started *Electrical World,* soon to supersede his *Operator,* he could claim that "there are men who got their first idea of a duplex from diagrams and explanations in *The Operator,* who now rank as first-class electricians, and draw pay as such." Although mobility within the industry had become much more restrictive by this time, he continued to argue that "to read intelligently and study assiduously is the only way to genuine success."[49]

An operator who called himself "Common Sense" also recognized the importance of demonstrating technical competence. In an 1878 letter to the editor of the *Journal* he argued that as "the higher positions in the telegraphic service, such as managers of important offices, chief operators, superintendents, etc., will be filled from the ranks, and those who exhibit special qualifications will have the best chance of filling vacancies in these positions as they occur . . . it will *pay* to qualify ourselves for them."[50] Telegraph superintendent John Abernethy gave operators similar advice in his manual, *The Modern Service of Commercial and Railway Telegraphy,* which first appeared in 1882. In a section devoted to the development of electrical engineering and the role of science in telegraphy, Abernethy wrote that "there is *always* a deficiency in the number of those who are properly qualified to fill the higher positions of honor in the telegraphic service, as circuit managers and the positions which require something more than the ability which is popularly supposed to suffice as mere mechanical manipulators and readers of telegraphic signals." Noting that new improvements were constantly being made in telegraphy, he told operators that "it is from the intelligent and ambitious class of employees that the offices of honor and profit in the telegraphic service of the future must mainly be filled."[51]

Operators also had before them examples of prominent company officials who had gained their positions through the mastery of technical details. George Prescott had risen from the operator's key to successive positions of responsibility, finally becoming company electrician of Western Union. Anson Stager was an operator in Cincinnati when he devised his important improvement in circuit design. He subsequently became superintendent of Western Union's western lines and was placed in charge of government telegraphs during the Civil War. After the war he became a major technical adviser to company president William Orton. John Van Horne, superintendent of the company's southern division, also began his career as an operator. He quickly earned the position of general manager for the Southwestern Telegraph Company under Norvin Green, who later appointed him vice-president in charge of Western Union's electrical and patent departments.[52]

Many less prominent individuals who served as chief operators, circuit managers, and office managers also received promotions based on their technical proficiency or on their managerial abilities.

Even without the advice of the journal editors, most inventors understood that a relationship existed among their personal ambitions, their inventive work, and the goals of telegraph companies. Most telegraph inventors began their careers as operators and thus were part of one of the first mass workforces to be employed by a large-scale corporation. This corporate context helped mold their aspirations. With the rise of the large-scale corporation, the ideal of success gradually shifted away from that of the heroic inventor seeking fame and fortune. Instead, invention became increasingly associated with more modest career aspirations.

Such career aspirations, however, were unavailable to the small but growing segment of female members of the telegraph community. Although women had served as operators in the United States since the beginning of commercial service in 1846, their numbers did not increase significantly until the Civil War, when companies actively recruited them to replace male operators joining the ranks of the military telegraph service. By 1870 there were approximately 350 women operators representing about 4 percent of all operators. Their numbers and percentage increased only gradually during the next decade and a half to about 1,400 in 1886, accounting for 7 percent of the workforce, but by 1900 there were over 7,000 women in the operating force, which amounted to 12 percent of the total.[53] The same economic forces that led middle-class women into teaching and clerical jobs also led them into telegraphy, which was generally considered a respectable lower-middle-class occupation. Although paid significantly less than men, women operators earned more than their female counterparts in most other occupations.[54]

Women who joined the operating force found admittance to the telegraph "fraternity" more difficult. Women often entered the industry in ways that challenged the traditional "apprenticeship" system by which experienced operators taught messenger boys or other young men interested in learning the profession. As Western Union president William Orton explained in outlining his plans for a company-sponsored telegraph school for women at Cooper Union in New York City,

> The supply of males is always abundant, being kept up by the opportunity which nearly every office affords for educating boys. There are but few of our offices at which it is practicable to educate women, and we have found that such education as has been afforded at seminaries and "commercial colleges" does not make practical operators.[55]

Women also challenged male operators in other ways. The initial influx of women during the war precipitated a debate in the pages of the National

Telegraph Union's *Telegrapher* over their suitability as operators. While the journal's editor looked favorably upon women operators and argued for their inclusion as members of the union, many male members objected to both their employment and their membership. Arguing that women were less skilled, they also questioned the suitability of telegraphing as an occupation for women, and at a time of declining pay, male operators expressed concern over the potential impact of low-waged female operators. Although some men remained hostile, women had generally become accepted members of the telegraph community by the mid-1870s, gaining respect for their abilities and for their support of strikes against Western Union and other telegraph companies. Furthermore, as their numbers grew it became apparent to the community that the low wages paid to women were a symptom of the larger economic forces causing declining wages rather than their cause.[56]

Western Union, in fact, generally used the lower wage scale for female operators to expand service. Noting that salaries accounted for about 60 percent of operating cost, Orton claimed that

> At many stations the gross receipts are but little more than the salary of the operator. It is one of our rules, however, to open and maintain offices at any point reached by our lines when the receipts will be sufficient to defray the expenses of such office.[57]

Thus, the company usually employed women in small rural offices and in small offices located in hotels and other public places where the number of messages would not justify a higher-salaried male operator. Increasingly, women operators also found employment in large metropolitan offices where they were usually employed on less heavily used branch lines rather than on main lines, although some did work the heavily trafficked circuits.[58]

Although they were part of the telegraph community and read the same telegraph journals as did male operators, women were unlikely to see articles about the relationship between invention and career mobility as having much significance for their more limited circumstances. "In a calling whose opportunities for mobility were fading, the prestige and economic rewards of managership were even more elusive for women than men."[59] Unlike their male counterparts, female operators were unlikely to be promoted on the basis of technical abilities. Their only route of advancement was to become a manager of other women operators in the metropolitan offices. Career mobility for women was further limited by the nineteenth century's dominant ideology of domesticity, which encouraged most women to leave the workplace for marriage. Although Western Union, unlike most European telegraph services, did not require women to quit after marrying, it was expected that most would do so and few women pursued a permanent career in telegraphy. The work culture of female operators may also have reinforced the era's dominant

Figure 3.11. By 1873 female operators had become common enough to be pictured in a *Harper's New Monthly Magazine* article. They were most often employed in small rural offices such as this one. (*Source: Harper's New Monthly Magazine* 47 [1873]: 332.)

domestic ideology. In her account of the debate over the employment of female operators, Melodie Andrews notes that "telegraphers came to terms with the presence of lady operators by creating within their profession a working-class variation of the domestic sphere for women."[60]

The expectation that female operators would exert an ennobling influence on the workplace is also evident in Edwin Gabler's study of American telegraphers. Some women even performed traditional domestic tasks in the operating room. An 1883 account of the Ladies' Department at Western Union headquarters described the women whiling away time between messages by "knitting, crocheting, or sewing" and noted that "conversation in a low tone is encouraged."[61] Although other evidence suggests that the workplace was usually not so free and easy, even on lightly trafficked wires, such domestic activities may well have been tolerated if not encouraged.

Technical expertise was even less expected of those few women found in telegraph manufacturing. Unlike operators, women employed in telegraph shops were denied access to traditionally male positions in the machine shop, especially operating machine tools. They were confined largely to the demanding, though repetitive, task of insulating wires and winding electromagnets, which may well have been perceived as requiring comparable skills to winding bobbins of thread.[62] Furthermore, in the male culture of the machine shop they were even less likely to find technical skill a source of promotion.

Not surprisingly, an examination of telegraph patents to 1890 only turned up two female telegraph inventors. The most notable of the two women was Bernice J. Noyes, who took out eight patents on fire alarm telegraphs between 1887 and 1890 and another seven in 1891. Several of these she assigned to the Municipal Signal Company, which used two of them as the basis of a patent infringement suit filed against the Gamewell Fire-Alarm Telegraph Company, the leading company in that field. Unfortunately, little is known of Noyes herself.[63] The other woman inventor was Clara M. Brinkerhoff, a music teacher who invented a telegraph key after becoming associated with former telegraph operator George Cumming. Cumming had devised a new design for a telegraph key, which Brinkerhoff and he improved by redesigning its contact points. The Cumming and Brinkerhoff patent on "periphery" contact points, granted at the same time as Cumming's telegraph key patent in 1882, was also applicable to other telegraph devices. They subsequently formed the firm of Cumming and Brinkerhoff to sell not only the key but also their periphery contact point design (which received several awards) as disc electrodes for general use in telegraphs.[64]

Manufacturing for general sale, a traditional method of introducing inventions into the marketplace, remained an important model for aspiring telegraph inventors. A number of operators such as Cumming used their inventive talents as a vehicle for entering a manufacturing career. Charles Chester had been one of the first to pursue this option in the early 1850s, and in the decade following the Civil War, it became an important path for ambitious operator-inventors such as Thomas Edison. This avenue proved particularly fruitful as the number of telegraph manufacturing firms expanded in the early 1870s to supply new markets for telegraphy in the cities as well as to meet the demand of new telegraph companies competing with Western Union. A large number of new telegraph manufactories were founded in major cities along the eastern seaboard, including New York, Philadelphia, Boston, Washington, D.C., Baltimore, and manufacturing cities in New Jersey. Others appeared in major inland centers such as Cleveland, Cincinnati, Chicago, and Indianapolis. San Francisco also had an important telegraph shop. These new shops were frequently established by individuals who were already members of the American telegraph community as operators or machinists.[65] Many of these shops, several of them supplying equipment for emerging markets providing communication within rather than between major cities and manufacturing towns, became the loci of inventive activity in the new urban telegraph industry that arose following the war.

One of the most successful operators to travel the path of inventor-manufacturer was Jesse H. Bunnell, who first attracted the attention of Philadelphia manufacturers Partrick and Carter with his improved repeater, patented in 1868. They manufactured the invention and in 1871 Bunnell became a member of the firm, which became known as Partrick, Bunnell and Company. After moving to New York City to open a branch office of the firm, Bunnell worked briefly with the large manufacturing shop of L. G. Tillotson before setting up J. H. Bunnell and Company, which became one of the largest electrical manufacturing concerns in the United States.[66]

Bunnell, like many other telegraph manufacturers, not only continued to invent but also encouraged and supported the work of other inventors, including his own machinists. Because successful manufacturing required the knowledge and experience of skilled mechanics, invention served as a means of upward mobility for machinists as well as operators. Sometimes a talented machinist became a business partner or superintendent. A few inventors, such as Elisha Gray and Thomas Edison, established partnerships with mechanics in which their associates ran the shop while they focused on technical improvements. Even though they operated on a relatively small scale, most telegraph shops required a substantial outlay of capital and only a limited number of manufacturers could expect to be profitable even in the largest cities.

While the path of inventor-manufacturer remained viable through the 1870s, few inventors found success following the traditional heroic model of the inventor-entrepreneur. In the 1840s and 1850s, inventors had been able to use patented new systems to compete effectively because the small scale of telegraph networks did not provide clearcut, nontechnical advantages, thus allowing claims of technical superiority to be tested in the market. By the end of the Civil War, however, Western Union's ability to provide truly national service made it difficult for even a technically superior system to better serve customers. Inventors working on new systems for the national telegraph network found themselves working directly for Western Union or one of its rivals.

Because invention increasingly occurred within a corporate rather than an entrepreneurial context operators began to view it as a means of moving up the corporate ladder and they aligned their career goals with those of the large companies for whom they worked. The rise of corporate engineering departments in the 1870s was closely related to changing corporate views of invention as corporate managers sought greater control over their large-scale organizations and encouraged greater technical efficiency. No longer concerned merely with securing the work of independent inventors, they developed bureaucratic structures that could take advantage of individuals in the corporation with technical expertise.

The career of Joseph Fenn illustrates the rise of corporate engineering staffs. Fenn became a telegraph operator during the Civil War, becoming an office manager for the Atlantic and Pacific Telegraph Company in 1869. He was an electrician for one of the short-lived stock printing companies that emerged in the early 1870s in New York City before joining the staff of Western Union electrician George Prescott in 1876. He assisted Elisha Gray in his experiments on a harmonic telegraph system and was also involved in the installation of the company's duplex and quadruplex telegraph systems. Fenn later became assistant electrician of the American Union Telegraph Company until it was taken over by Western Union. Returning to Western Union, he served for a time as assistant circuit manager before becoming electrician of the company's southern division. Fenn himself took out a number of duplex and quadruplex inventions in the 1880s.[67]

The country's railway systems with their own bureaucratic organizations provided another alternative route of advancement. Although the nation's rail lines were integrated and coordinated through cooperative arrangements each company had its own set of superintendents and managers. Because no one railroad company dominated employment, as Western Union did in commercial telegraphy, railway telegraphers formed the Association of Railway Telegraph Superintendents in 1882, which remained the only national organization of telegraph operating officials. Although the majority of telegraph operators were located in railroad offices, they performed many nontelegraphic duties and few acquired the same proficiency as did commercial operators.[68] Those who did take the time to improve their telegraphic skill and knowledge could achieve promotion. Some managers and superintendents also came to railroad telegraphy after experience in commercial offices.[69] The career of Charles Jones illustrates how this might work. While an operator in Western Union's Albany office in 1868 Jones patented a switchboard, which he offered to the company for five hundred dollars. The invention was adopted and came into common use, thereby enhancing Jones's reputation, and by the early 1870s he had become superintendent of telegraphs for the Illinois Central Railroad. In 1882 he served as the first secretary of the Association of Railway Telegraph Superintendents.[70]

Although the opportunities for advancement in the industry declined by the early 1880s as the number of telegraph companies declined, thus reducing the number of offices and management positions, operators seeking advancement could and still did use invention as a means of demonstrating important abilities to their employers. Daniel Grandy was a Western Union operator in New York and Boston for twelve years when he patented a duplex relay design in 1880. That same year he was made a manager in the St. Louis office. Grandy, whose career was set back when he participated in the 1883 strike, eventually became assistant chief of the company's southern division.

THE JONES PATENT

LOCK SWITCH.

For simplicity and durability this improved Switch has no equal. The brass bars over which the current passes *are in plain view.* The connection formed by the plug is PERFECTION ITSELF. The method of locking is simple and effective. *The plug cannot become loose and drop out.* There is *no back board to warp and split,* and *no hidden wires and connections to become unsoldered by Lightning.* It requires but little space in an office.

In fact, it is the most perfect, and the lowest priced first-class Switch yet invented.

These Switches are made with or without Lightning Arresters, and any desired size.

PRICES:

With Lightning Arrester attached, $3 00 per perpedicular Bar.

Without Lightning Arrester $2 50 per bar for Switches from four to thirty bars (two to 15 lines).

Figure 3.12. This advertisement from the catalogue of telegraph manufacturers L. G. Tillotson & Co. describes the advantages of Charles Jones's patent switch. (*Source:* L. G. Tillotson & Co., *Price List of Telegraph Machinery and Supplies* [New York: privately printed, n.d.], 74.)

William Athearn, who became an operator in 1877, received his first patent for a telegraph key in 1882. A year later he received his second patent for a repeater, and in 1885 his last patent for a duplex telegraph. During these years he worked for several telegraph companies and advanced to more responsible positions as chief operator and office manager. By 1890 Athearn had become chief night operator at Western Union and four years later was appointed electrician of the operating department.[71]

For Athearn and his generation of operators the predominant model of success had become upward mobility within the confines of the corporate bureaucracy. Although this changing pattern was becoming visible during the 1870s, its impact was initially blunted by new opportunities opening up in fast-growing urban telegraph markets. Besides providing new jobs for technically competent operators, new urban telegraph services also spurred an increase in telegraph manufacturing and encouraged the concentration of the industry's technical community in larger cities. The small scale of local telegraph systems also revived the possibility for inventor-entrepreneurs to use new inventions to enter a market, although industrial consolidation would again reduce such opportunities in the 1880s. The rapid technical development of urban telegraph systems also spilled over into the long-distance market as some of these same inventors were encouraged to devise new technology intended to provide competitive advantages to companies competing with Western Union.

4

The Urban Technical Community
and Telegraph Design

I N THE DECADE following the Civil War, American telegraphy took on
an urban character that greatly influenced the kind of work inventors
undertook, the manner in which they developed their ideas, and the
kind of support they received. During the industry's first decade technical
personnel and corporate organizations were spread out among the many
small regional systems. Even though most operators continued to work in
rural and small-town offices during the next decade, the consolidation of
telegraph companies, the growing dominance of larger cities as business and
communications centers, and the establishment of telegraph manufactories
in them began to concentrate the industry's business and technical leaders
in urban areas. This trend toward urban concentration was accelerated following the Civil War.

Among the factors contributing to concentration was the market structure of long-distance telegraph communication. After the war, companies
seeking to compete with Western Union's national system faced extensive
line-construction costs and needed access to investment bankers and other
capitalists, who were concentrated in New England and the middle Atlantic
states, particularly Boston, Philadelphia, and New York City.[1] Telegraph companies had their headquarters in these same cities, with New York domi-
nating the industry, but Boston and Philadelphia continued to be important
centers. A few other cities, such as Chicago in the Midwest and San Francisco
on the Pacific coast, became minor centers by serving as regional headquarters for Western Union's administrative divisions and by virtue of their role
as regional market centers. Not surprisingly, the more creative members of
the technical community tended to work in these urban areas in order to
gain access to telegraph entrepreneurs and company officials who might
support their work.

The growing proximity of telegraph inventors to urban businessmen in
turn encouraged inventions that became the basis of a variety of specialized
telegraph services, including market report, private line, and messenger and

alarm services. Because the telegraph reinforced the advantages held by the larger cities as centers of commerce, it was in these cities that commodity exchanges arose in coordination with the telegraph, which also made possible the centralization of securities markets by the New York Stock Exchange. Private telegraph lines, which could coordinate a firm's internal communication, and market-reporting instruments, which helped coordinate external business information, enabled firms to take greater advantage of such centralized management. Instruments for facilitating such intelligence gathering were a principal focus of telegraph invention in the decade following the Civil War.[2] The growth of urban telegraph markets created new support for telegraph inventors at a time when long-distance companies, especially industry leader Western Union, paid little attention to innovation. This would change as new long-distance companies emerging to compete with Western Union sought technological advantage by turning to the creative members of the urban telegraph community for new inventions.

Urban telegraph systems also influenced telegraph design and the manner in which inventors developed their ideas. Because urban telegraphs were intended to be reliable and easy to use, inventors often focused their attention on the mechanical aspects of design. Their experience working on urban telegraphs also influenced the design of new long-distance telegraph systems. The predominance of electromechanical design considerations in telegraphy reinforced the importance of manufacturing shops in the inventive process, at the same time that the growing markets for urban telegraph instruments in turn encouraged the establishment of new shops. These shops provided telegraph inventors with crucial resources of knowledge and skill as well as employment and financing; together with the growing telegraph markets, in the nation's cities and manufacturing towns, they helped to bring about the urbanization of the industry's technical community.

The urbanization of the telegraph community also reflected trends causing the quickening pace of invention throughout American society following the Civil War.[3] Invention in the United States increasingly became an urban activity, closely aligned with a growing industrial base and urban population. Though many inventors first gained experience with machinery on farms and in small village shops, cities and manufacturing towns provided important resources necessary to the successful inventor.[4] Nathan Rosenberg has suggested that a confluence of metalworking industries using similar techniques gave rise to a learning process that was important in the evolution of machine tool design.[5] Others have speculated that the critical flow of information fostering such a learning process and generating innovation was best nurtured in the urban environment. More explicitly, Steven Lubar has shown how the textile industry of Lowell, Massachusetts, created a community of textile machinery inventors. Lubar has also shown how exchanges of infor-

Figure 4.1. Large operating rooms such as the one at Western Union headquarters in New York City were important centers for the urban telegraph community. (*Source: Harper's New Monthly Magazine* 47 [1873]: 335.)

mation and skill among machinists and inventors at the Hoe shops in New York affected not only printing press technology, but also influenced the design of pinmaking machinery. Telegraph invention also proved to be a cooperative enterprise in which a technical community, largely concentrated in urban centers, provided key resources to inventors.[6]

The Urban Telegraph Community and the Process of Invention

Though the majority of telegraph inventors lived and worked in small towns scattered along the telegraph lines,[7] the careers of the leading telegraph inventors reflected the importance of major American cities in cultivating and motivating technical creativity. The most successful inventors lived in or near large urban centers, such as New York.[8] These cities offered the inventor a substantial technical community located in the large main operating rooms and in the precision machine shops of telegraph manufacturers. In the operating room and the machine shop, technically sophisticated telegraph inventors exchanged information and studied telegraph apparatus. The cities also provided access to local capitalists willing to fund invention as a speculative enterprise or to form new companies on the strength of an invention. The location of telegraph company headquarters in the large cities also meant access to company officials who might support their company's purchase of an invention or even support new inventive work. In the course of their labors inventors found it necessary to call upon the wide range of technical and financial skills offered by the urban setting.

The importance of the resources offered by the urban technical community is suggested by the experience of David Flanery. As the Western Union division superintendent at New Orleans, he invented a printing telegraph that he offered to the company and its subsidiary, the Gold and Stock Telegraph Company, in 1873. Because of his isolation from the center of printing-telegraph technology, his design was not equal to the state-of-the-art instruments already owned by the company. In explaining why the invention was not purchased for service with the Gold and Stock business, Western Union president William Orton noted "the genius, scientific knowledge and mechanical skill" that Flanery's invention evidenced but suggested that he was hampered by not

> having had opportunity to come in contact with inventors of telegraphic apparatus and to be benefitted by the stimulus which such contact affords . . . And it does not reflect upon the possessor of these that others first in the field and enjoying better opportunities appear to have fully occupied it for the present.[9]

Flanery kept in touch with the larger technical community through the pages of technical journals,[10] and presumably through current books on the subject of telegraph technology, but he had limited access to other technically sophisticated members of the community who congregated in the major northern cities. Like most other operators, those in New Orleans enjoyed a certain public prestige for presiding over the efficient operation of telegraph lines and offices. Typically, most of them possessed only a limited comprehension of the workings of the apparatus. Furthermore, New Orleans did not have a manufacturing community specializing in telegraph and electrical instruments.

In contrast to Flanery, Thomas Edison, whose printing telegraphs Gold and Stock was using, worked in major manufacturing cities and his inventive efforts were stimulated by the technical communities he found in them. Though Edison's major success as an inventor came in New York, his earliest successful inventions were made in Boston. Soon after arriving in this leading center of telegraph invention and manufacturing Edison made it a point to visit the city's manufacturing and electrical shops. He described these shops and their products in an article for the *Telegrapher* and also wrote other articles describing the items they manufactured. Recognizing the importance of shop facilities for his ambitions as an inventor, he sought to become familiar with the local manufacturing community. Edison found space in a corner of one shop for experiments and, after acquiring support from a local entrepreneur, set up a small shop of his own with machinists formerly employed by other Boston telegraph manufacturers. His investigations of manufacturers' instruments also spurred his inventive work. One of the machinists that worked with Edison, George Anders, had built the dial telegraph manufactured by Edmands and Hamblet, whose shop and instrument Edison described at length in his article. Together with Anders, Edison went on to invent and market his own dial telegraph.[11] Edison was one of many inventors who found in Boston's telegraph shops a crucial source of knowledge and skill.

Samuel Bishop, a prominent New York manufacturer of insulated telegraph cables, explicitly acknowledged the central place of urban machine shops for the exchange of knowledge within the telegraph community when he established a "telegraphic exchange" in 1868. The exchange consisted of space in his shop where instruments might be displayed and inventors and others interested in telegraph improvements could "be brought together oftener, where views can be freely interchanged, which will ultimately enure to the benefit of the public by cheapening telegraphic transmission."[12] Although this attempt to provide an informational forum was short-lived, shops such as Bishop's did provide opportunities for more informal exchanges as evidenced by Edison's experience in Boston. Published descriptions of other manufacturing centers also indicate a willingness on the part of manufacturers to allow technically competent visitors to examine instruments and undertake experiments. Describing David Brooks Paraffine Insulator Works in Philadelphia, with its laboratory and testing room, the *Telegrapher* noted that "telegraphers of a scientific and investigating turn of mind cannot fail to derive pleasure and instruction from a visit to this well appointed establishment."[13]

The precision machine shops of telegraph manufacturers also provided the primary workplace for the would-be inventor. In Boston, for example, the telegraph manufacturing shop of Charles Williams, Jr., contained a laboratory used by prominent telegraph inventor Moses Farmer, who also had his instruments constructed there. It was there that Edison first found space for his experiments. Many other telegraph inventors, including Alexander Graham Bell, went to Williams's shop to have their experimental instruments built and tested. A number of shops employed mechanics who specialized in working on experimental devices for inventors. As already noted, Thomas Watson made experimental apparatus for several inventors, including Bell, while working at Williams's shop. In Chicago, Ernest Warner found that "new inventions [were] being frequently placed in my hands, by inventors, to construct under their own supervision, with the consent of the Western

Figure 4.2. Charles Williams, Jr.'s shop in Boston was noted for its skilled experimental machinists and support for inventors. (*Source:* AT&T Archives, Neg. 2751.)

Electric Manufacturing Company," for which he worked. As further assistance to inventors, Western Electric itself had a laboratory for "electrical, chemical, and other scientific investigations."[14]

Inventors in small towns and isolated railway stations were seriously hampered by their lack of access to a machine shop. They either built patent office models or experimental devices themselves (or perhaps had the local blacksmith or jeweler do so) or took their drawings to larger towns where manufacturing shops were available. For a brief period in 1875, Ezra Gilliland's shop in New York allowed inventors to send drawings from which devices could be assembled out of his large stock of parts, thus saving the cost of having special orders made, but this was not a typical practice.[15]

The importance of the machine shop to inventive enterprise was recognized by a number of successful inventors who engaged in manufacturing. Besides the income it provided, the shop owner also had the advantages of a readily available machine shop at no extra cost. In his own shop, the inventor did not have to wait in line for service. Edison experienced this problem as a young inventor soon after he moved from Boston to New York. Writing to one of his financial backers in Boston, he explained that his lack

Figure 4.3. The workmen of Gray and Barton in Chicago, later Western Electric Manufacturing Company, with Elisha Gray's popular private-line printer, which had been designed and manufactured in the shop. (*Source:* AT&T Archives, Neg. 88-200609.)

of progress in conducting experiments was due to "awaiting the alteration of my instruments which on account of piling up of jobs at the instrument makers have been delayed and I will probably have to wait one week longer."[16] Edison made it a point to include provisions in his later contracts for inventive work that enabled him to set up his own manufacturing shops and to support machine shops at his laboratories.

Other inventors also recognized that the optimum design of a new instrument depended on continuous access to the resources of a machine shop. While inventing his improved transmitter for printing telegraphs, Charles Wiley was often at the manufactory of Pearce and Jones, working by the side of the mechanic making his instrument. Wiley would provide the mechanic with sketches and verbal instructions, suggesting changes in the construction of the device.

The close interaction that often took place between inventor and machinist challenged the traditional view that true invention was a flash of insight. Few inventions sprang full-grown from the brains of their inventors; their births usually required skilled mechanics acting as midwives. Mechanic Paul Seiler's testimony in a patent interference case suggests the potential prob-

Figure 4.4. The precision machine shop at Thomas Edison's Menlo Park laboratory suggests what a telegraph manufacturing shop probably looked like. The equipment in the laboratory had originally been installed at Edison's Newark manufacturing shops. (*Source:* ENHS, Neg. 6935.)

lems this collaboration could cause in the patent office, but also illustrates why the traditional view remained dominant. Seiler had been approached by George Ladd, president of the Electrical Construction and Maintenance Company that employed Seiler as foreman, and also an officer in the American District Telegraph Company of San Francisco, about designing a new call box for American District. In describing their collaboration, Seiler had difficulty remembering who had contributed what to the design that Seiler later patented on his own after Ladd decided not to pursue the project. Indeed, he considered it to be a "kind of a mutual property." But Seiler's subsequent testimony suggested that his own ideas about the intellectual property of invention favored the person, in this case Ladd, who originated the idea for the invention, even if he did not reduce it to practice. Seiler went on to say that

> if I make a model for a man the idea may be his, and yet if I make a model and suggest different things I call it mine. It is mine, not by lawful or legal right, but I call it mine. I may not be able to patent the idea but mechanics are generally apt to manipulate these things.[17]

As a result, even though the interference testimony indicated that it was Seiler and not Ladd who reduced the idea to practice, Seiler himself believed that since the original idea for the design embodied in his patent had probably come from Ladd, that Ladd and not he was the inventor.

Even after the machinist built a device, experiments often suggested further changes necessary to improve its operation. The pattern of constructing, testing, and altering a machine was an important part of the creative effort of invention.[18] The way in which an inventor used such experimental devices was elaborated upon by George Phelps while testifying in a patent interference case:

> Experiments indicate probable practical results that are to be obtained by perfected details. They may necessitate the manufacture of several pairs of instruments to produce them in the best practical form. I have never known experimental first instruments to be in condition, though demonstrating the correctness of its theory for practical service, it requiring time and patience to construct and arrange its mechanical details for practical use.[19]

As important as experimental results might be, actual operation of apparatus was essential to developing the best combination of mechanical and electrical operations for a new invention. While most manufacturers had test lines on which inventors could experiment, they frequently found that there was a

> great difference between the experimental working of inventions in a laboratory and in practical operation upon telegraph lines. The conditions are so different, that it is not unusual to find that apparatus which appears to be all right when experimentally tested, is impracticable or defective in operation when tried in actual business upon long lines.[20]

Thus, Charles Buell used Western Union lines for experiments with his multiplex fire and burglar alarm system in order "to determine proper proportions and adjustments on circuits of considerable length."[21] Orazio Lugo, experimenting with dynamos to replace batteries for telegraphy, acquired permission to use a loop line of the American Union Telegraph Company extending from New York to Boston, with a view to convincing the company's officers "that with one single machine lines of very great length could be worked while shorter lines could be operated at the same time."[22]

Only a trial on actual working lines could allow such a claim to be tested and inventors needed access to company officials in order to convince them to allow such tests. Permission to experiment on existing lines was most easily obtained in the larger cities where certain office managers and superintendents were well known for their interest in such experiments. In Cincinnati during 1867, railroad telegraph superintendent Charles Summers conducted experiments on duplex telegraphy with operator Thomas Edison. Later, as electrician of Western Union's central division, he assisted Henry Nicholson and Francis Jones in their quadruplex experiments. At Boston, where Edison decided to actively begin his career as an inventor in 1868, George Milliken, the Western Union office manager, was a noted inventor actively engaged in experimental work with Moses Farmer. Less prominent office employees aided in other ways. For example, chief operator Emile Shape of Western Union's Milwaukee office loaned inventor Charles Haskins a condenser for line tests of his duplex system. Other operators helped test instruments over the lines.[23]

Even with such assistance, invention was a costly enterprise. Patent assignment records suggest that inventors often found support for their work from local capitalists as well as from company officials who provided funds for experimental and patent costs.[24] In addition, the assignment records indicate that urban centers provided the most ready access to these sources of support. Urban centers also provided access to the resources of manufacturing shops, where inventors found skilled machinists to build and alter instruments and test lines on which to conduct experiments and the occasional manufacturer who was willing to assist them by extending credit.

For many inventors the time and expense of experiments prevented them from transforming their ideas into patentable inventions until they could acquire the necessary capital. Manufacturers often provided the initial assistance an inventor required to turn his idea into a practical device he could use to attract investors or take before a company official. For instance, while Charles Wiley was working on his printing telegraph inventions in the 1870s, he kept a running account with the manufacturing firm of Pearce and Jones, whom he found to be "very liberal in the matter of money with me."[25] Henry Van Hoevenbergh established a close relationship with manufacturer Jesse Bunnell, who gave the inventor "the freedom of my shop, to a certain extent, that he might easily have had small models . . . completed without being brought directly to my knowledge."[26]

Some manufacturers provided such help because of friendship. Often, however, they sought their own competitive advantage. In his study of early manufacturing in the telephone industry, George Smith observed that Bell Company managers assumed that "competitive variations in equipment design [were] useful to the development of the technology."[27] The company initially employed several shops, all of them formerly devoted to telegraph manufacture, to construct its instruments. Very likely, management assumptions about technological development derived from the experience of the telegraph industry. The evidence of advertisements and catalogs, and the willingness of manufacturers to enter their instruments in industrial fairs

Figure 4.5. The variety of patented inventions advertised during the 1870s in journals such as the *Telegrapher* indicates their significance in competition between telegraph manufacturers. (*Source: Telegrapher* 9 [1873]: vi.)

attests to the importance they attached to differentiating their apparatus from competitors. While not all improvements were patented and some changes were superficial, many were intended to enhance a firm's reputation for quality. Manufacturers often sought to patent these improvements or gain control of them through patent assignments, often acquiring an exclusive manufacturing contract as well. Western Electric, for example, established a policy of acquiring patents relating to electrical apparatus to use as a competitive tool.[28]

To protect an invention, it was necessary to obtain a patent. The patent also provided a legal title that the inventor could sell in the marketplace. Patenting an invention was expensive, however, and inventors sometimes put off applying for a patent because of a lack of funds. Inventors faced a series of fees required by the U.S. Patent Office, and the intricacies of patent law usually required the services of patent solicitors. In addition to the attorney's fees, inventors paid fifteen dollars to file an application, which had to be accompanied by a model and standardized drawings, both of which required additional expense. Once the application was allowed, the inventor was required to pay an additional fee of twenty dollars before it could be issued. Further costs could be incurred if the patent was put in interference with a competing application or issued patent.[29] The experience of William B. Watkins, who found himself without funds either to have models constructed or to pay the final fees on his patent applications for fire alarm telegraphs, was not unusual. Even a relatively successful telegraph inventor, such as Thomas Edison in the mid-1870s, sometimes lacked funds to pay final fees and allowed several months to elapse before having patents issued.[30]

Although inventors sometimes found support from family members for their inventive work,[31] the early career of Thomas Edison, who relied on local capitalists and corporate officials located in the larger cities, was more typical. Indeed, Edison would later draw on the experience of his early career when advising young inventors to apply to local capitalists or company superintendents when trying to find support for their first inventions.[32] In giving such advice he was describing a pattern common to many telegraph inventors of his generation.

Edison began his career as an inventor while working as an operator in the main Boston office of Western Union in 1868.[33] Among his inventions was a stockbroker's printing telegraph that he assigned to fellow operator Dewitt Roberts on the condition that Roberts "furnish or cause to be furnished sufficient money to patent and manufacture" one or more of the instruments. Roberts was apparently unsuccessful and Edison turned for support to John Lane, a stockbroker and former president of the Franklin Telegraph Company. Lane, too, failed to provide sufficient funds for Edison's experiments and the inventor found new investors, Joel Hills and William

Plummer, two local businessmen who not only advanced costs for patents and experiments, but who also established a stock quotation business in Boston using Edison's instrument.

Edison found backing for other inventions from E. Baker Welch, a director of Franklin Telegraph, who was typical of the local urban capitalists who hoped to turn other people's inventions into fortunes. Among the Edison inventions in which Welch had an interest were a double transmitter, fire and burglar alarm telegraphs, and a dial telegraph. With Welch's backing Edison established a business providing private line telegraphs using the dial instrument and also set up a small manufacturing operation. Welch also provided support for Edison's friend and fellow operator Milton Adams by advancing experimental costs for his printing telegraph, and to George Anders, the telegraph machinist who worked in Edison's shop.[34] Using Welch's money, Edison traveled to New York in the spring of 1869. There he became associated with companies providing stock (and other) market reporting services with printing telegraphs. Edison soon established himself as a key inventor of printing telegraphs and, through the officers of these urban telegraph companies, he became associated with officials at Western Union and other intercity telegraph companies.[35]

The continued growth of urban telegraph services during the 1870s created new interest in the work of inventors among local entrepreneurs, such as those supporting Edison's work in Boston. Entrepreneurs in other major cities and large towns provided similar support for inventors as they sought to enter thriving new markets for stock and commodity reports, private line telegraphs, fire and burglar alarms, and messenger services. With nearly half of all patented telegraph inventions during the 1870s and 1880s devoted to urban telegraph systems, it is not surprising that most telegraph inventors congregated in the major cities and manufacturing towns of the United States.[36]

Invention and Design

The design considerations of the new urban telegraph systems reinforced the electromechanical character of telegraph invention and the role of the machine shop in inventive activity. As urban telegraph inventors and promoters sought to make their systems readily accessible to a wide range of users, they attempted to reduce the need for skilled operation. Most inventors therefore attempted to design machines simple to operate and requiring limited electrical know-how. Mechanical design was often more important than electrical design in ensuring reliability.[37] These design factors helped to make the urban telegraph shop the central institution of telegraph invention during the 1870s, by reinforcing the influence of mechanical technology and American machine shop practice on telegraph invention. By providing a source of inventive talent upon which long-distance firms could draw, the urban telegraph community also influenced the design of long-distance telegraph systems as well.

The earliest urban telegraph services were private business lines using the same equipment as intercity telegraphs and requiring the employment of skilled Morse operators.[38] The first such lines were set up around 1850 by telegraph superintendents and by a few large businesses able to afford the relatively high salaries of skilled Morse operators.[39] Users outside of the telegraph industry began to install private lines in significant numbers only with the introduction of instruments that, even if initially more expensive, saved labor costs. The first significant designs that allowed private lines to be operated without skilled Morse operators were dial telegraphs, which had clock-like faces displaying letters and numbers marked by a moving indicator. Although invented by Charles Wheatstone in England in 1840 and used widely in Europe on long-distance lines, dial instruments were little used in America until the 1860s when they were adapted to urban lines. By the end of the decade a growing market for intracity telegraph communication led a number of inventors, many of them manufacturers, to turn their attention to this potentially lucrative market.

In designing dial telegraphs, inventors sought to make them usable by those unskilled in telegraphy. This is evident in Edison's description of Edmands and Hamblet's dial instrument

> In operating this instrument no knowledge of the usual Telegraphic signs or sounds are necessary; the operator simply places his fingers upon the letters of the alphabet which compose the Telegram, and the person receiving simply takes notice of the letters as they are successively pointed out upon the indicator at the other terminus.[40]

The Edmands and Hamblet instrument also contained another design feature common to dial telegraphs—it obviated the need for batteries, which required careful maintenance that involved messy and often dangerous chemicals. Indeed, one of the hindrances to consumer use of electrical technologies such as a private-line telegraphs was the need to replenish the chemical batteries. Dial telegraph inventors therefore focused on designs that used induced electric currents generated when a soft iron armature was revolved between the poles of a permanent magnet. Edison's own dial instrument, the "magnetograph," was named for the magnetoelectric source of its current. Although easy to use, dial instruments were hampered by their relative slowness. Edison's co-inventor, George Anders, increased the speed of his later dial telegraph design by employing a keyboard transmitter.

ANDERS'
Magneto Alphabetical Dial Telegraph Instruments

Are the best Dial Instruments ever invented. Having more powerful Generators than any others, they produce stronger electrical currents, and will therefore work on longer lines and with larger dials and indicators. The reading of messages is much easier than with the smaller dials and indicators used on other instruments of this kind. For the same reason they are more RELIABLE in bad weather, or when the state of the atmosphere is unfavorable to instruments having Generators of less power.

THE CITY OF BOSTON USES THESE INSTRUMENTS ON ITS POLICE TELEGRAPH.

EDWARD H. SAVAGE, ESQ., Chief of Police, in his annual report made in January, 1874, says :

"We have now Anders' Magneto Dial Machines, which are arranged with a simple keyboard and a circular alphabetical indicator, and the whole can be easily operated by a child of ten years, if he has learned to spell. There is a machine at the central office, and one in each station, and each of the eleven stations is connected with the central office by an independent wire. *The whole now gives such entire satisfaction that we believe we have as simply-constructed, as simply-worked, and as useful a Police Telegraph, as can be found in the world.*

☞ Each Instrument has a Signal Bell to call the person who is to receive the message. ☜

Figure 4.6. An advertising brochure for George Anders's magneto-dial telegraph instrument noted that it had a more powerful generator than others, allowing it to be used on longer lines as well as making it more reliable in bad weather. The instrument also had larger dials, making it easier to read messages. (*Source: TAEM* 27:203.)

ANDERS'
Magneto Type Printing Telegraph Instruments.

For Manufacturers, Coal and Lumber Dealers, Mining and Gas Companies; Railroad and Police Telegraphs, &c., &c.

PATENTED IN THE UNITED STATES AND EUROPE.

ANY PERSON CAN WORK THEM! NO ACID OR CHEMICAL BATTERY USED!

The electrical currents are generated by induction from permanent magnets, and as the messages are spelled out on the keys, they are printed on a strip of paper by each instrument on the line. The printing is done as rapidly as any person can select the letters on the keys. For reliability, rapidity, economy, and convenience of operating, combined, they are superior to any other printing telegraph instruments yet produced, either in Europe or the United States.

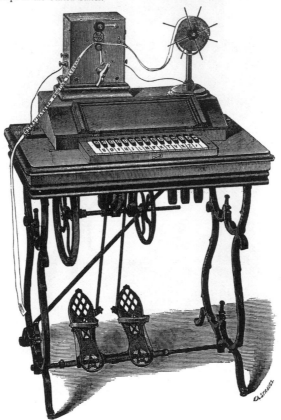

Figure 4.7. This advertisement for Anders's magneto printer shows its similarities to his magneto-dial instrument. The design of the table and footpedals derived from those used in foot-powered machine tools and sewing machines. (*Source: TAEM* 27:202.)

The real growth of private lines began in the early 1870s with the development of small, inexpensive printing telegraphs.[41] Like dial instruments, printing telegraphs did not require a skilled Morse operator, but they could also transmit a message with no receiving operator at all. Initially printers used dial instruments as transmitters, but inventors soon adapted the more rapid keyboard design of long-distance printing instruments. Small printing telegraphs spread rapidly, and several leading telegraph inventors refined them.[42]

Printing telegraphs did have one important drawback for most users: they required batteries. The one printing telegraph that employed magneto-electric currents rather than batteries was the one designed by George Anders. Unfortunately, Anders's instrument was also more costly. Thus, George Ladd wrote Anders's business partner E. Baker Welch that "out of our experience we believe a Magneto printer properly constructed is better than an electro-magnetic printer. But yours is too expensive for our market."[43] Other potential customers also indicated that the Anders machine, which sold for $200 to $250, was too expensive when other printing telegraphs, most notably those of Thomas Edison and Elisha Gray, sold for between $125 and $150. Nonetheless, there was a significant market for the machine because it was reliable and did not require batteries. As one potential customer put it "*none of the other styles are safe to be left alone*" (emphasis in original).[44]

Although printing telegraphs had been used on long-distance lines for two decades, the private-line printing telegraphs were an outgrowth of the most characteristically American development in urban telegraphy, the stock ticker. The ticker had its genesis in 1866 on the New York Gold Exchange, whose enterprising vice-president, Samuel Laws, devised an electrically operated indicator that showed the current price of gold to the trading floor and, through a street window, to the outside world. In 1867, small indicators on a single telegraphic circuit were placed in the nearby offices of brokers and merchants. At the end of that same year, the first printing stock reporting telegraph, invented by telegraph operator Edward Calahan, was introduced, replacing the runners who had previously carried information from the trading floor to brokers' offices.

Calahan devised his stock ticker by replacing the dials of an indicator with two typewheels—one for letters and the other for numbers. Calahan was able to simplify the design of his printer by taking advantage of the relatively low costs for stringing wires in dense urban areas. He used extra line wires to operate the simple escapement mechanisms that controlled the printing mechanism. His printer thus operated with three line wires—two to rotate and ink the typewheels and a third to print and advance the paper. Calahan's system also initially used local batteries for each instrument as was common in private-line and other telegraph systems, but brokers objected

Figures 4.8 and 4.9. Samuel Laws originally devised a dial to transmit gold prices to the floor of the exchange. Later he invented a small dial indicator for use in brokers' offices which could also be used for transmitting the price of stocks, bonds, produce, and other commodities. His patent (U.S. Pat. 75,775) also indicates that he intended to develop an instrument that could print out prices as well as indicate them on the dials. (*Sources:* Medbery, *Men and Mysteries of Wall Street,* 231; Cat. 551:40, ENHS.)

Figures 4.10 and 4.11. Samuel Laws's stock printer, designed to compete with Edward Calahan's invention, continued to incorporate a dial until it was radically altered by Thomas Edison. Edison greatly simplified and improved the printer's design by reducing the instrument's size and number of moving parts as well as its electromechanical operation. Although no evidence exists as to why Edison chose a frame similar to those found in contemporary sewing machines, he may have been influenced by the instrument's unknown manufacturer. (*Source:* Cat. 551:5, 14, ENHS.)

Figure 4.12. Edward Calahan designed the first stock "ticker." In place of indicating dials he used two typewheels—one for letters and the other for numbers—to print permanent messages on strips of paper. His original printer used three wires: two to rotate and ink the typewheels and a third to operate the printing mechanism and advance the paper. (*Source:* Cat. 551:2, ENHS.)

to the damage caused to carpets and furniture when the sulfuric acid solution in the battery was renewed. He therefore devised a central battery system that obviated the need for local batteries and became standard for market-reporting systems.

Almost immediately after their introduction stock tickers became indispensable to Wall Street and the focus of furious inventive and entrepreneurial activity. Within a few years there were reporting instruments on circuits based at the cotton, produce, petroleum, and other commodity exchanges in every major American city. Inventors working on private-line and market-reporting printing telegraphs faced similar technical requirements. In fact, Edison's private-line printer was an adaptation of his most successful stock printer. However, market reporting, especially from the Stock Exchange, also presented special problems for the designers of printing telegraphs. The machines needed to be small, fast, reliable, and reasonably inexpensive. Mechanical design improvements proved to be an important means for achieving these goals. For example, the operation of the two typewheels used in most machines presented several opportunities for mechanical improvements. Inventors thus introduced design improvements such as escapements that allowed the typewheels to rotate more rapidly and devices to shift between the two typewheels in order to reduce the number of wires needed to operate the machines or to achieve greater speed and reliability.

Synchronizing all the machines on the line so that they printed the same information created a special technical problem. Unlike private-line systems, which linked two instruments for point-to-point communication, market-reporting printers on a line were operated by a single transmitter at the exchange. Because printers frequently fell behind the transmitter by one or more letters, making the message useless to the customer, exchange companies initially sent employees to the office where the printer was running "out of unison" to reset it. The frequency with which printers fell out of unison led to the development of automatic mechanical means to solve the problem. One of the most effective and longest used devices was Edison's screw-thread unison in which a peg seated in a screw thread on the typewheel shaft allowed the transmitting operator to bring all the printers on a line into unison by sending sufficient impulses to turn the shaft until the peg hit a stop. The relationship between electrical circuits and mechanical design of typewheels, unisons, and printing mechanisms was crucial to the speed and reliability of a machine, but mechanical considerations remained the critical factor.[45]

Edward Calahan also originated another line of technological development when he devised the district telegraph system in 1871. Before Calahan's system, private entrepreneurs providing telegraph systems to facilitate business and social communication used the Morse equipment and trained operators to send and receive messages. These early systems were largely unsuccessful and independent city telegraphs did not become common in the United States until the 1870s, following the introduction of Calahan's district telegraph system.[46]

The Calahan district signal box employed two push-keys with which the customer signaled either a messenger or police or fire service. Subsequent designs also used push-keys or dials, and many of these allowed customers to signal for other services such as doctors and carriages. The mechanism of the signal box included breakwheels with several notches cut into them. As these breakwheels turned, they caused a number of electrical impulses to be sent, thus signaling the number of the box and the desired service. Falling weights, springs, and clockworks were used to drive the breakwheels. The greater the number of signals required by a box the more complex was its

Figure 4.13. Edison's screw-thread unison was a major improvement for synchronizing stock printers. In the unison a peg seated in a screw-thread on the typewheel shaft allowed the transmitting operator to bring all the printers on a line into unison by sending sufficient impulses to turn the shaft until the peg hit a stop. (*Source:* Acc. 29.1980.1313, from the collections of the Henry Ford Museum and Greenfield Village.)

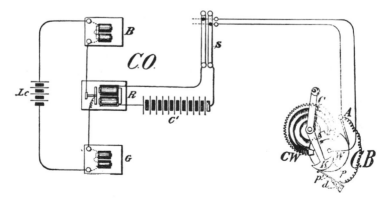

Figure 4.14. A basic district telegraph circuit showing the mechanisms of a simple messenger call box. The notches in the breakwheel W caused the circuit to break as the wheel turned and they made contact with the contact spring A. In this example, there are three notches together followed by a single notch. This would indicate to the central office that the signal came from subscriber box 31. (*Source:* Maver, *American Telegraphy,* 368–69.)

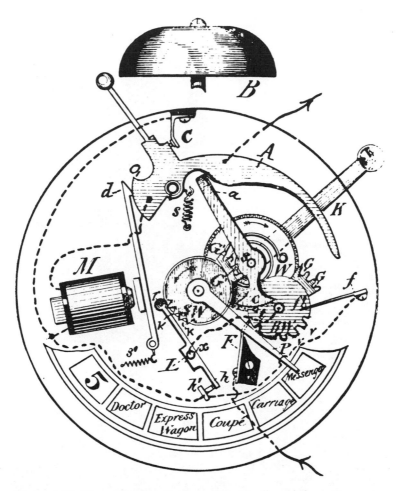

Figure 4.15. More complex district call boxes allowed the customer to call for several different services and also allowed the central office to acknowledge the message by sending a return signal that rang a small bell. In this device the pointer caused cogs *GG* on the breakwheel *BW* to engage one or more of the cogs *G'G'* on a second breakwheel *SW*, sending the appropriate number of signals to indicate that particular service. (*Source:* Maver, *American Telegraphy,* 374–75.)

mechanical design. Like stock-ticker circuits, district telegraph systems were powered by batteries located at the central station, where receiving instruments, similar to Morse registers or automatic telegraph receivers, were used to record the coded messages. Messenger boys stationed at the offices then responded to subscribers' signals for police, firemen, transportation, the family doctor, or messenger and other services.[47]

District telegraph companies established their offices in urban and subur-

Figure 4.16. Double pen registers such as this were commonly used by both district and fire-alarm telegraph companies. Each pen was operated by a separate clockwork mechanism activated by the closing of an electromagnet. In this way one pen was always kept ready for operation. (*Source:* Maver, *American Telegraphy*, 372.)

ban neighborhoods and placed the small automatic transmitting devices in subscribers' homes and businesses. The local companies also delivered messages for and to the urban offices of the national telegraph companies, an arrangement that helped stimulate the further development of the district systems. By the end of the decade, many of the neighborhood district offices also served as the first central exchange offices for the newly introduced telephone.[48] While messenger departments remained the principal part of the district telegraph companies' operations, the role of the district telegraph was further enlarged by the development of nightwatch services. Companies established their own private police patrols, which quickly became popular and profitable. They later developed extensive private fire alarm services as well. By the end of the century, as the telephone eliminated the need for messenger services, the district companies turned primarily to providing protection systems and a variety of new inventions were developed for these services, which evolved independently of the telegraph industry.[49]

In designing his original district telegraph call box, Edward Calahan used already existing municipal fire alarm systems as his model. As in the district signal box, the fire alarm box used notched breakwheels that transmitted a coded message when a crank was turned or a lever pulled. Both systems also had similar technical requirements, especially ease of use and the ability for two or more boxes to be worked simultaneously without interfering with each other. Although circuit designs and methods of testing for grounded lines were important to their smooth operation, mechanical elements were the primary consideration in designing transmitters and receivers. Because of this, John Ruddick, a largely illiterate and untutored mechanic with little knowledge of electricity, was able to make a major improvement in fire-alarm signal boxes as late as 1888. In that year, Ruddick devised the first signal box that did not interfere with others operating on the same circuit at the same time. Although he could not have made the invention without the limited knowledge of electricity he picked up after coming to work for the Richmond Fire Alarm Company, Ruddick's mechanical skills provided the key to solving the problem.[50]

The basic design used in signal boxes was invented by William Channing and Moses Farmer, who installed the first fire-alarm telegraph in Boston in 1851. Although few fire-alarm telegraphs were installed before the Civil War, during the 1870s they became a fixture in American cities. The Channing-Farmer system was widely adopted and improved by several inventors, but others competed by devising new systems. By the mid-1870s seventy-nine cities in the United States and Canada had municipal systems in combination with professional municipal fire departments; at the turn of the century they could be found in over six hundred cities and towns.[51]

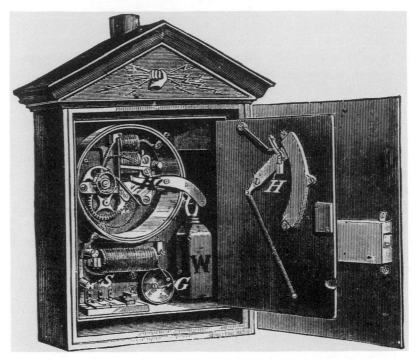

Figure 4.17. Weight-driven fire-alarm signal boxes used a series of notched breakwheels to signal the location of the box in similar fashion to those used in district telegraphs. (*Source:* Maver, *American Telegraphy,* 439.)

Fire-alarm telegraphs revolutionized firefighting, and their success was not lost on police officials. By the early 1870s, police-call systems were being established in a few major cities, many of them by the same company marketing the Channing-Farmer system of municipal fire alarms. Initially, large police departments adopted systems employing dial transmitters and printing receivers such as those used by urban businesses. For example, Charles Chester offered a combined fire-alarm and police telegraph system that incorporated his dial telegraph design for police calls, while George Anders convinced the Boston police department to use his magneto dial and printing telegraphs. But it was not until 1880, when John Barrett, then superintendent of the Chicago fire-alarm telegraph, devised a combined telegraph and telephone police call, that such systems became common.[52] While the telegraph message allowed the patrolman to indicate his location (and allowed his supervisors to keep closer tabs on his rounds), the telephone allowed officers to communicate necessary information rapidly and completely when asking for assistance, or when the station requested them to assist at another location.

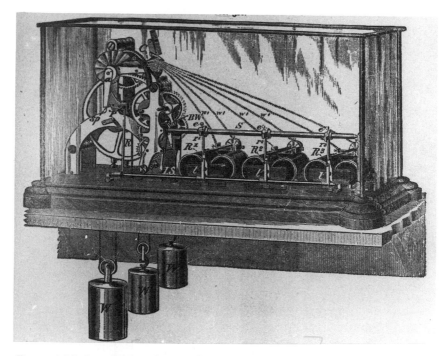

Figure 4.18. In order to relay an alarm signal from a central station to fire stations throughout the city, the Gamewell Fire-Alarm Telegraph Company introduced a noninterfering automatic repeater. The weight-driven gears on the left controlled a series of rods that locked the armatures of all relays except that of the original signal until the message had been repeated. (*Source:* Maver, *American Telegraphy,* 439, 441.)

Although the telephone's early use was in supplementing or replacing other systems of urban telegraphy, it grew out of attempts to devise a system of sending several messages simultaneously over a single wire on long-distance lines. The telephone emerged during experiments with what were known as acoustic or harmonic telegraphs in which a number of tuning forks or reeds transmitted different tones or frequencies over a wire. The major problem confronting inventors working on these systems was finding a suitable means for separating the signals at the receiving end. The telephone was born when one of the inventors working on such a system, Alexander Graham Bell, recognized that he could transmit and receive the human voice. Although telephony would evolve as a separate communications system that challenged long-distance as well as urban telegraph systems, it was initially perceived as a form of telegraphy.

Most urban telegraph services evolved independently of the long-distance telegraph lines, although companies providing intercity service did

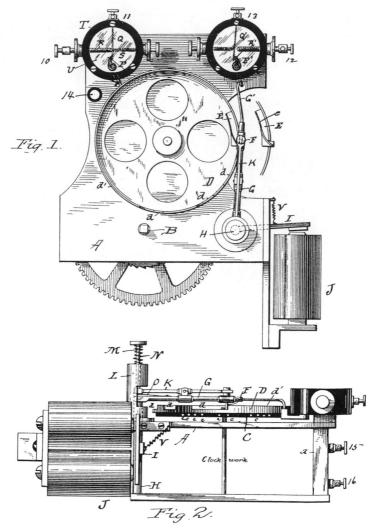

Figure 4.19. A patent drawing (U.S. Pat. 404,438) shows the operation of John Ruddick's noninterfering fire alarm, which mechanically delayed the sending of a signal by more than one box at a time. In normal operation, when the button M was pushed it would disengage pin F from plate E, allowing wheel D to rotate freely so that the numberwheel C could send the signal. When two boxes were operated at the same time, the breaking of the line circuit by the first box would cause the circuit-controller Q'R' of the second box to cut out magnet J, thus causing the armature to release and pin F to engage wheel D, preventing it from rotating until the first box finished signaling and the line circuit could again activate magnet J of the second box and allow its armature to close and disengage pin F.

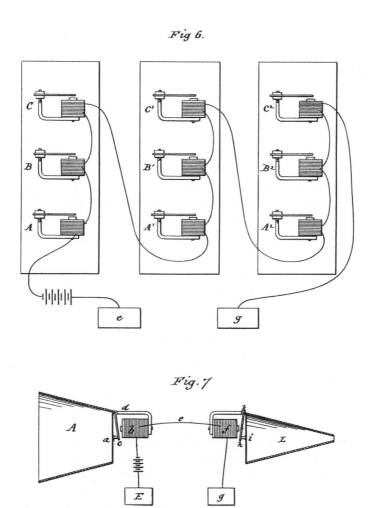

Figure 4.20. Alexander Graham Bell's acoustic telegraph (U.S. Pat. 174,465) included the basic principle of his telephone. The upper drawing shows the tuned reeds used in his acoustic telegraph and the bottom drawing is his magnetoelectric telephone.

find it advantageous to interconnect with local district and market-reporting telegraphs because they generated long-distance messages. More significantly, the technical know-how that developed the urban telegraph systems was applied to long-distance telegraph technology and helped to create a period of rapid technical development during the 1870s. While electricity provided its motive force and presented unique technical problems on long lines, long-distance telegraph technology also incorporated mechanical devices and knowledge of mechanical movements remained valuable to inventors. Indeed, the electromechanical aspect of telegraph invention was so prevalent that James Reid felt compelled, in his 1879 book on the telegraph industry, to point out that telegraph inventor Elisha Gray's outstanding "characteristic as an inventor is in avoiding mere mechanical devices to produce results . . . He seeks to make electricity do its work direct, and therefore endeavors in his devices to train and harness it for that purpose."[53] Yet, among the many important inventions made by Gray were several printing telegraph designs that relied on mechanical aptitude as well as electrical knowledge. Other prominent inventors working on both urban and long-distance telegraph technology, such as Thomas Edison and George Phelps, were referred to as "electro-mechanicians."[54] Indeed, the elite of telegraph inventors enjoyed proficiency in both facets of telegraph design. Their combination of electrical and mechanical ingenuity gave them great versatility, allowing them to work on a variety of telegraph systems.

An important aspect of Edison's great creativity was his facility with mechanical design. His mechanical fecundity was a product of long study as well as native genius; for example, in 1872 he filled a notebook with over one hundred escapement designs.[55] This dictionary of escapements included many of the standard designs reproduced in such works as Munn and Company's guide for inventors, but Edison added his own variations. His notebooks from this period are filled with experimental telegraph instruments in which such escapements play an important role. Edison's draft of a book on telegraphy, in which he planned to present the results of his many years and experiments in the field, suggests the importance he placed on electromechanical considerations. One chapter (of which there are several extensive drafts) was devoted almost entirely to mechanical movements used in breakwheels, and another focused on printing telegraphs, whose designs were extensively influenced by their mechanical elements. Even chapters on the basic electrical actions of different telegraph systems and devices were titled "Electrical Movements" and "Miscellaneous Movements," suggesting that Edison may have conceptualized these as mechanical analogs.[56]

Edison's work on long-distance telegraph technology provides insights into the electromechanical nature of telegraph technology. Even problems requiring extensive electrical knowledge for their solution, such as multiple

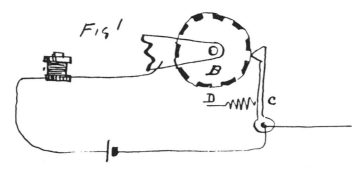

Figure 4.21. Edison's draft of his chapter on breakwheels shows the most common form used in telegraph apparatus. (*Source: TAEM* 7:253.)

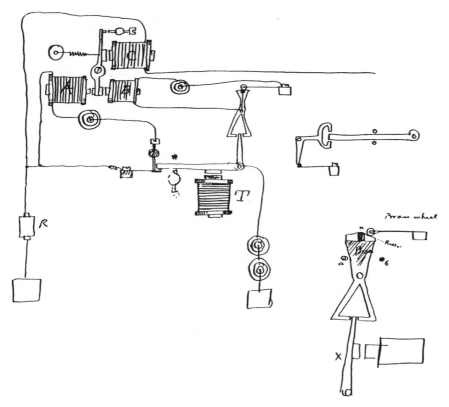

Figure 4.22. Edison's drawing for a model of one of his quadruplex designs with an electromechanical compensator (enlarged in detail), which was activated by the sounder lever striking against the triangular anchor piece and causing the brass center at the top to make and break contact with the roller, thus neutralizing induced currents that would otherwise alter the line balance. (*Source: TAEM* 10:323–24.)

telegraphy, provided scope for Edison's mechanical ingenuity. Several experimental designs for one of his most famous and important inventions, the quadruplex telegraph system, which allowed four messages to be sent simultaneously over one wire, involved solutions such as mechanical compensators used to balance the electrical forces acting on the armature of a receiving magnet and electromechanical methods of isolating interference from simultaneous signals. Edison also experimented with mechanical solutions for acoustic telegraphy.[57]

Edison's other major contribution to long-distance telegraphy, a system of automatic telegraphy, challenged him mechanically as well as electrically. In his attempt to increase the speed of the Bain system of automatic chemical-recording telegraphy, Edison found that the inertia of the mechanism used in his instruments limited the speed with which the signal could be transmitted. In attacking the most difficult problem of high-speed automatic telegraphs, an electrical phenomenon that distorted the signal, Edison turned to mechanical as well as electrical solutions. This effect, known as "tailing," resulted when induced currents, set up when the circuit was broken and the line discharged, caused a prolongation of the signals. This effect was imperceptible at moderate speeds, but caused an elongation of the recorded message at high speeds or long distances. As a result, dots often appeared as dashes, while dashes might cause the entire message to appear as a solid line. Edison's first attempt to solve this problem involved devising a transmitter that mechanically switched in a battery to reverse the polarity of the line current, thus discharging the wire.[58] Other methods of improving signal transmission involved various combinations of perforations in the transmitting tape to improve signal transmission. In order to change the perforations, Edison needed to redesign his mechanical perforating machines.[59] Mechanical design also challenged Edison when devising methods of driving his transmitters and receivers. Although the instruments actually used on the line were hand-cranked, he also devised clockwork mechanisms similar to those found in his district telegraph receiver.[60]

Nonetheless, as Edison's successful solutions to transmission problems in quadruplex and automatic telegraphy demonstrated, electrical knowledge was an essential part of telegraph invention. Gray spoke specifically to the differences between local and long-distance telegraphy in a letter to his patent solicitor, A. L. Hayes, regarding reports that Bell's system of acoustic telegraphy was suitable for sending voice transmissions. Gray explained why he had not pursued his own experiments, suggesting the possibility of transmitting the human voice.

Many instruments work splendidly on short circuits that utterly fail on long ones. A long circuit has several additional conditions to be taken

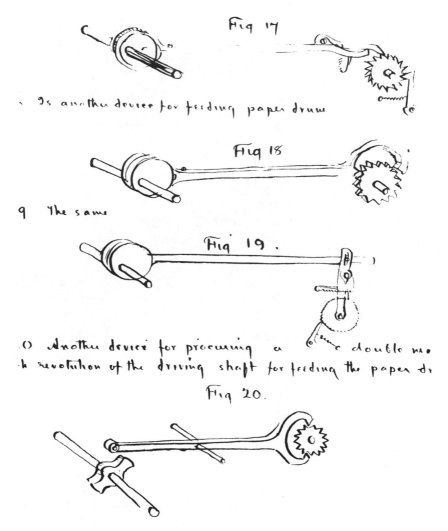

Figure 4.23. Edison conceived of a variety of escapement designs to control the paper-feed mechanism for one of his automatic telegraph instruments. (*Source: TAEM* 3:83.)

care of besides mere resistance. My long and large experience in practical matters pertaining to telegraphing enables me to work understandingly. I therefore reject at a glance many things that would seem to one unacquainted with the freaks of long line phenomena, something very valuable.[61]

Gray's statement was an accurate appraisal of the limited range of early telephone technology, but it also suggests that he ignored his own experience

Figure 4.24. This photograph shows the clockwork mechanism for Edison's electrochemical recording instrument used in his district and fire telegraphs. (*Source:* Cat. 6895, ENHS.)

with urban telegraph systems in appraising the potential value of such an invention.

Both Edison and Gray focused their early inventive work immediately after the Civil War on key problems in long-distance telegraphy—self-adjusting relays and repeaters.[62] Yet, both inventors, as well as many of their contemporaries, found their initial success in urban telegraph markets. Because urban telegraph systems competed in a very technologically competitive environment, several companies devised strategies aimed at controlling the technology and its inventors. As companies in the long-distance market, including Western Union, began to adopt new technology they also found it necessary to evolve corporate strategies toward invention.

5

Invention and
Corporate Strategies

ORPORATE officials conceived their attempts to establish greater
control of technology primarily in market terms, but they were also
influenced by their own beliefs regarding the technical and business
characteristics of telegraphy. Control of technology could be used for a variety of purposes: an existing company could maintain or expand its market
share, or enter new markets, while a new company might use technology to
enter an existing market or to create a new one. Firms competing with
Western Union in the long-distance market required technology that would
reduce their line construction costs and give them technical advantages over
Western Union's ability to provide national service. During the 1870s, when
telegraph technology seemed to be undergoing rapid development, Western
Union officials found it advantageous to pay close attention to new inventions
in order to prevent such technology from falling into the hands of competitors.

The ability to use technology for competitive purposes differed with the
nature of the technical systems and the markets in which they competed.
The large scale of long-distance telegraphy gave Western Union significant
competitive advantages with its existing national system and the failure of
competitors to develop radically new technology subsequently led to reduced
support for invention in this market. Indeed, Western Union's most successful competitors found technology of only limited use. Maintaining dominance
in any of the urban telegraph markets was much more difficult because of
the relatively small scale of local telegraph systems. As a result, entry costs
were low and competitors could bid for customers by offering lower prices,
better instruments, or better service.

In both long-distance and local markets, the key to competitive strategies
involving technology proved to be access to inventors or their patents. Certain patents were critical to the technology in a particular field, and firms
controlling them could assume a dominant position through the filing of
patent infringement suits. The resolution of complicated technical issues by

a court of law, however, often depended on the quality of legal talent hired by the parties involved, and a competing firm willing to hire such talent could defend itself against such challenges. Even a successful infringement suit might provide only short-term protection because of the limited life of a patent. Thus, companies acquired direct control of the work of important inventors in a particular field in order to assure control of key patents. They also supported inventive work designed to reduce the costs of instruments, increase their reliability, or improve operating efficiency.

Urban Telegraphy and the Rise of Corporate Strategies

Urban telegraph markets remained generally competitive throughout the nineteenth century, although certain markets did come to be dominated by national firms. The most competitive markets were those made up of numerous small users rather than large institutions. Such markets were usually more competitive, and companies providing reliable, cheap instruments had an important advantage over their rivals. Competition often focused on the relative merits of the instruments provided by competing firms. Private-line and district telegraph services were two markets that remained generally competitive.

The local emphasis in district telegraphy was a product of the original strategy adopted by the American District Telegraph Company of New York, which controlled the original Calahan patents on district signals. American District of New York licensed these and other patents it owned to independent companies in other cities, most of which also became known as American District Telegraph. Because of this licensing practice, neither American District of New York nor other local district companies was inclined to pursue competitive strategies based on control of technology. Instead, they competed on quality and cost of service and generally only sought new inventions that could improve efficiency and reliability. Occasionally they also adopted inventions that could be used in providing new services, such as fire or burglar alarms.

Although competition in district telegraphy continued to take place primarily at the local level, it was greatly affected by interconnection with major long-distance telegraph companies.[1] New competition in this market was generated when Western Union established close ties to many of the original local American District Telegraph Companies. This led rivals to establish ties to competing companies or to establish their own district services.[2] Because district systems were less costly to build and patent infringement was not as important a consideration, they attracted local entrepreneurs, such as those who used Thomas Edison's patents to establish the Domestic Telegraph Company in 1874. In order to compete with American District Companies offering long-distance telegraph service through Western Union, Domestic Telegraph soon found it advantageous to interconnect with Western Union's principal competitor at the time, the Atlantic and Pacific Telegraph Company. Atlantic and Pacific, of course, was hoping such interconnection would help it to gain the same advantages accruing to Western Union from its association with many of the local American District Companies.[3] Other Western Union competitors also established connections with existing district companies or used rival inventions in order to establish competing service and gain the advantages of interconnection. At the same time, local district companies used competition between national companies to acquire more advantageous terms of interconnection.[4]

Nonetheless, competition among long-distance telegraph companies for connections with district telegraph systems did not necessarily translate into support for new technology. This is illustrated by the experience of John Wilson who, as the American Union Telegraph Company's chief operator in Boston, had charge of that company's district telegraph system as well. In 1880 American Union, which had become Western Union's main competitor, introduced a new district signal system into Boston in its attempt to compete with Western Union, which controlled the local American District Telegraph Company. Wilson told company officials that he could design a cheaper, simpler, and more reliable signal box, but found little support for his work from American Union, which was apparently satisfied that its existing box served the purposes of providing an alternative to American District. Wilson therefore used his own resources to conduct experiments. He had his first experimental signal box built by a Boston watchmaker, whom he paid directly for the instrument, and his second and third boxes were constructed by a prominent local gear-cutter in exchange for a half interest in the invention. When Western Union took over American Union in 1881, Wilson was given charge of a Western Union branch office in Boston and was at the same time engaged by American District. Though he earned a salary of one hundred dollars per month from this dual employment, which was equal to that of a first-class telegraph operator, Wilson found himself unable to pay for experimental boxes he was having constructed at Charles Williams's shop.[5] Failing to convince American District officers to invest in his experiments, Wilson turned first to American District superintendent Henry Poole and then to local entrepreneurs to support his work and to establish a company to introduce his invention.[6]

Wilson was unable to gain corporate support for his improvement because the Boston market for district telegraphs did not make a better signal box design crucial to maintaining or increasing market share. In other circumstances, however, district companies might well attempt to maintain dominance through control of technology. Thus, the businessmen who owned rights to Paul Seiler's district telegraph box patent pursued the patent interference, even though Seiler himself refused to be a party to it, because they were using it in competition with George Ladd's American District Telegraph Company of San Francisco.[7]

In contrast to district telegraphy, the markets for business news, generally furnished under the auspices of stock exchanges and boards of trade, and for municipal fire and police alarms, which were the province of incorporated cities, quickly became dominated by single firms. These firms used combined strategies of controlling patents and securing license agreements with responsible institutions to emerge as virtual monopolies in their field, although competitors were able to use new technology to enter these markets and, in some cases, to compete rather effectively with the dominant national firm on a local level.

The Gamewell Fire-Alarm Telegraph Company, which evolved an early strategy of technical innovation and superior service as a means of achieving a near monopoly in its field, exemplifies this pattern. The company's founder, John Gamewell, initiated this strategy following the Civil War when he sought "to have every patent issued on Fire Telegraph examined as soon as issued and if found to contain any point of value at once to purchase the same and incorporate it into [his] system."[8] Gamewell did more than just examine patents being issued for fire alarms; he made important inventions himself and also encouraged others to make improvements in the company's system.

Because mechanical design made the machine shop a central institution in the development of urban telegraph systems, Gamewell turned to manufacturer Moses Crane when his company experienced problems with its automatic signal boxes in the late 1860s. Crane, one of the principal manufacturers of the company's apparatus, was requested to assist company operator Edwin Rogers with his suggested improvements. Crane and Rogers became joint inventors and later made other important contributions improving the company's system. Crane's factory became the center for much of the inventive work done for the Gamewell Company.[9]

Although the Gamewell system was considered the best by experts and in actual trial, competitors could successfully underbid it. City officials interested in re-election sought ways to save taxpayers' money when providing new services, and price as well as reliability became an important selling point in fire alarm telegraphy. To maintain its reputation and to prevent competition, Gamewell Telegraph continually sought improvements by its own inventors and by others, and initiated interference proceedings or filed infringement suits against rival systems with the ultimate goal of gaining control of the patents. If a competing system was adopted by a municipality, Gamewell Fire might take legal action against the city, but it diplomatically refrained from seeking an injunction against continued use of the infringing patents until the city was properly protected by the Gamewell system. By the end of the century, Gamewell Fire could claim that it had "furnished more than ninety per cent" of all the municipal fire alarms in the United States and Canada, attributing its success to a policy that sought "not only to extend and promote the introduction of Fire-Alarm and Police Telegraphs, but to improve and perfect their service."[10] The experience of the Gamewell Fire-Alarm Telegraph Company may well have provided a model for other telegraph companies, although there is no direct evidence that it did so.

The emergence of similar strategies at the Gold and Stock Telegraph Company had more obvious repercussions within the telegraph industry because they affected the strategy of Western Union. Gold and Stock, using Edward Calahan's stock ticker, began operations in 1867 as a local firm serving only the New York Stock Exchange. The company soon expanded into other financial markets and acquired its principal rival, Samuel Laws's Gold Reporting Telegraph, in 1869.[11] By the following year, Gold and Stock, under the direction of its new president Marshall Lefferts, established a policy "to control all improvements which may present themselves and by the possession of which we can more surely control our business, perfect our system, extend its usefulness, and render it the most complete in the world."[12] Toward this end he reviewed the company's existing contracts with Calahan to ensure the company's control of them, acquired important patent rights held by the Laws company, purchased the printing telegraph patents of Franklin Pope and Thomas Edison, and secured Edison's services as a contract inventor. Lefferts also secured exclusive leases with both the Gold and the Stock Exchanges.[13]

Gold and Stock's support for Edison illustrates the importance of the machine shop to telegraph invention. Even before acquiring the patents Edison and Pope were using in their private-line and gold reporting business, two officers of Gold and Stock engaged Edison's services. Their contracts for Edison's work on a facsimile telegraph and a stock printer included provisions that made Edison a consulting electrician and provided funds that enabled him to set up a small shop that became his first major manufacturing firm, Edison and Unger. Edison's manufacturing facilities also provided the resources for his inventive work for Gold and Stock, which continued to support Edison and Unger and its successor, Edison and Murray, through manufacturing contracts for Edison's instruments.[14]

As Lefferts moved aggressively to put his policy into effect, he also challenged Western Union's financial news services when he increased his company's capital stock in order to begin extending its business to other commodity exchanges and other cities throughout the United States.[15] Western Union was then making its own commitment to private-line and market-

Figure 5.1. The Gold and Stock Telegraph Company provided funds for Thomas Edison's first manufacturing shop in Newark, N.J. This photograph shows Edison's employees and two of his instruments that they manufactured for Gold and Stock. Note the resemblance to the photograph of Gray and Barton workmen in Figure 4.3. (*Source:* ENHS, Neg. 6047B.)

reporting telegraphy, under the direction of president William Orton, who believed that "if this field must be occupied, the ability to control it should remain with us [rather] than to permit an alliance to be made with the opposition." The national company was already providing a similar service with its commercial news department, formerly managed by Lefferts. Orton recognized that the growing network of urban systems could provide an important captive market for intercity telegraph messages. As a result, he supported the inventive efforts of George Phelps, superintendent of the Western Union factory, to develop an improved printing telegraph and convinced Western Union's board of directors to purchase Phelps's patent for $20,000. Seeking to obtain the advantages of interconnection with Gold and Stock, Orton also negotiated with Lefferts, leading to a consolidation of the two companies.[16] His expectations were realized after the takeover, when Gold and Stock contributed a "large increase in messages transmitted by the wires of the Western Union Company."[17]

Under Western Union's control Gold and Stock continued to actively encourage inventors in printing telegraphy and to seek control of new patents. As a result of the consolidation, Gold and Stock controlled the services of a number of key inventors, including George Phelps and Thomas Edison. Recognizing that Edison's ability to make seemingly endless improvements would help enable it to cover the field of printing telegraphy, the company entered into a new five-year contract with Edison for his services as consulting electrician and inventor. It also gained access to the services of Elisha Gray, electrician and inventor for the Western Electric Manufacturing Company, recently acquired by Western Union to strengthen its position in the urban telegraph market. As with Edison, Gold and Stock supported Gray's manufacturing firm through exclusive contracts to manufacture his printing telegraph inventions. Over the next two decades, Gold and Stock encouraged inventive work by its superintendents, acquired significant patents from other inventors, and aggressively pursued patent infringement suits when rival firms sought to use competing inventions to enter the field. It also sought to buy out its competitors in order to control their patents.[18]

Western Union's competitors, recognizing that interconnection with integrated systems providing urban communications services offered the industry giant important competitive advantages, sought to establish their own companies to compete in these markets. In the field of market reporting, for example, the Atlantic and Pacific Telegraph Company purchased the inventions of Henry Van Hoevenbergh, a former Gold and Stock employee, while the Mutual Union Company supported the work of John Wright and John Longstreet, the company's electricians, in their efforts to develop a printing telegraph system. Mutual Union also hired Stephen Field as a consulting electrician for this purpose. Field's inventions subsequently became the basis for Gold and Stock's most formidable competitor, the New York Quotation Company. New York Quotation was owned by the New York Stock Exchange, which had bought out the Commercial Telegraph Company, which owned Field's fast printer. Although Gold and Stock superintendent George B. Scott had devised his own fast printer, the Stock Exchange decided to own and control the service provided to its members. As a result, Gold and Stock was only able to transmit from the Stock Exchange to the offices of nonmembers. Nonetheless, at other exchanges in New York and other large cities Gold and Stock continued to be a dominant firm, aided by its connection with Western Union.[19]

During the 1870s, the growing network of urban telegraphs created a communications infrastructure that influenced the technical operations of companies providing long-distance, intercity telegraph services. Western Union, in particular, became aware not only of the potential competitive advantages of interconnection, but also of the use of new technology as a competitive tool. The concentration of the industry's technical community in urban centers, particularly New York City, had been encouraged by the tremendous growth of urban telegraph systems. When urban telegraph inventors began to apply their know-how to long-distance telegraph technology, they helped to create a period of rapid technical development. As Western Union encountered a more technically competitive marketplace in the 1870s, its interconnection with Gold and Stock provided the company's managers with crucial experience that helped them to forge new strategies toward invention and gain new influence over the work of this community.

The Development of Western Union's Strategies

Because Western Union emerged from the Civil War with a near monopoly over intercity message service, Western Union's managers focused their market strategy on maintaining dominance in the telegraph industry. Western Union's managers initially focused their attention on "minor" technical improvements in the company's existing system, to consolidate the myriad lines acquired from former competitors and improve the quality of service to western cities. Western Union's rivals, on the other hand, needed technological advantages to compete with its national network and service. They sought improvements that could substantially increase productivity or reduce costs while at the same time using far fewer lines. Such improvements also became attractive to Western Union's managers in the face of renewed competition in the 1870s.

Both Western Union and its rivals desired inventions that would increase the speed of message transmission or the capacity of a single telegraph wire to carry messages. Technical development involving line capacity and transmission speeds focused primarily on two types of systems. One system, multiple telegraphy, could be adapted to the existing Morse system and was designed to increase the number of messages that could be transmitted over a single wire. Automatic telegraphy, on the other hand, was intended to replace the Morse system with automatic machinery that transmitted at higher speeds than the Morse and also used less skilled, and hence, less costly operators. In determining which inventions to adopt and which inventors to support, telegraph officials' decisions reflected a set of conceptions and assumptions about the nature of telegraph business and technology. This "business-technology mindset" mirrored cultural values concerning the role of the telegraph and the nature of invention.[20]

The principal architect of Western Union's strategy in regard to invention was William Orton. Orton's initial introduction to telegraph technology came while he was a student at the State Normal College in Oswego, New York. There he wrote his thesis, accompanied by a model of his own construction, on the new electromagnetic telegraph that began commercial operations the same year that he graduated. However, little in Orton's subsequent career suggested that twenty years later he would rise to the top of the telegraph industry. While teaching school he also became associated with the Geneva, New York, publishing firm Derby and Company, becoming a partner in 1852. Orton continued in the publishing industry for many years, establishing the firm Miller, Orton, and Company in New York City in 1856. After the failure of this firm he worked for a while as managing clerk for the booksellers Gregory & Company. Orton's ambitions led him to begin legal studies and he also became active in city politics, winning election to the common council in 1861. Soon thereafter President Lincoln appointed him collector of internal revenue for New York, in which position he attracted the attention of Secretary of the Interior Samuel Chase. In early 1865 he was appointed U.S. Internal Revenue Commissioner, but seeing new opportunities in the telegraph industry Orton resigned in October to become president of the new United States Telegraph Company. Within a short time Orton discovered the company's precarious competitive position and decided to enter into negotiations with Western Union. The resulting consolidation made Orton Western Union vice-president in 1866 and he succeeded to the presidency the following year.[21]

When he became president of United States Telegraph, Orton lacked familiarity with the industry, but he quickly gained crucial insights into the business of telegraphy which were to be the basis of his management of Western Union. He came to believe that the best way to improve his company's competitive position would be to increase the volume of business with a minimum of additional expense. He saw that this could best be accomplished by improving the condition of the company's wires, thereby increasing their capacity, and by increasing the number of new lines operated by the company, thus attracting additional business.[22] As president of Western Union, Orton often returned to these themes. For example, in a letter of 4 December 1868, addressed to T. H. Wilson, a Western Union stockholder, he stated that "our policy concerning [opposition lines] is to put up more wires and otherwise constantly increase our facilities. In short, to grow faster than they and at the same time to do our business better."[23] Orton perceived that quality of service was more important than price to most Western Union customers, especially businessmen.

Early in his career at Western Union, Orton elaborated his definition of quality service, arguing that time was the company's real competitor. By taking all possible steps to provide the most rapid service, he would cause "the customers of our competitors to come to us for the protection of their own interests."[24] Therefore, it was in Western Union's interest "to adopt every improvement whereby the dispatch of business within a given time can

Figure 5.2. As president of Western Union, William Orton was the most powerful figure in the telegraph industry. The personal interest he took in technical decisions was a major factor in Western Union's growing support for inventors. (*Source:* Reid, *Telegraph in America,* 518.)

be materially increased."[25] Furthermore, he argued, "it is certainly cheaper for us to provide new instruments at almost any cost which will ever be charged therefor, than to put up, keep in repair, and operate additional wires to produce the same results."[26] Initially, however, Orton's willingness to adopt new machinery did not lead him to develop a strategy for obtaining such needed improvements. Instead, he waited on the initiative of independent inventors. Most firms interested in acquiring technical improvements adhered to this passive approach, because they considered invention to be a product of individual inspiration, which could be guided in only the most general fashion. It was up to the inventors themselves to direct their efforts toward particular markets in the hope they might gain fame and fortune.

Only slowly did companies such as Western Union perceive the value of attempting to direct inventors' efforts.[27]

Under Orton's leadership, Western Union did begin coordinating the way in which new technologies were investigated and adopted. In 1867 Orton hired British telegraph engineer Cromwell F. Varley to investigate the condition of the company's lines. This decision grew out of Orton's desire to increase the capacity of existing wires and ensure that new construction was of the best quality for prolonged service. Varley's report recommended that the company conduct regular electrical tests to insure the standardization of its wire, the quality of its batteries and insulation, and the resistance of its relays. The English electrical expert also suggested that the company evaluate available insulators and determine which were best suited for use on its lines. Following Varley's suggestions, Western Union undertook an extensive program of line reconstruction and improvement and began an extensive investigation of insulators.[28]

At the time these investigations began, no set procedure existed within the company for evaluating new or existing devices. As the need arose, individuals within the corporation who had special expertise would be called upon to give their opinions.[29] In deciding which insulator to adopt, Orton turned to his division superintendents, on whose lines the existing insulators had been used and tested. These men not only offered Orton expertise gained by direct experience but could make the necessary electrical tests and evaluations.[30]

The ad hoc nature of the company's approach to invention and the need for standardization in its technical operation eventually prompted Orton to reorganize the company's technical administration. In the fall of 1870, in conjunction with a general administrative rationalization, he established the office of electrician under George Prescott, a leading member of the technical community. Prescott's office came to play an important role in the commercial introduction of new inventions acquired by Western Union. Initially, the staff did not engage in basic research and development, which was still considered the province of the independent inventor. But the electrician's office did play a key support role in the work of inventors hired by the company, and members of the staff increasingly made important inventions of their own.[31]

Although Orton established an institutional framework for evaluating any improvements the company might adopt, he made the final decisions about new inventions, and when confronted with conflicting opinions from his experts, he trusted to his own judgment. Orton's study of the general laws of electricity and his practice of soliciting pertinent opinion and information strengthened his belief that, however much he had to learn and "imperfect" as were many of his impressions, it was not presumptuous of him "to hold and express opinions upon some technicalities of the service, as well as in respect of the policy which should govern the administration of the whole."[32] An industry journal obituary of Orton noted that the Western Union president personally supervised all aspects of the company's business and that he studied and gained intimate knowledge of the technology employed on its lines.[33] Orton's perception of the technical features of the telegraph service was an essential component of Western Union's strategy and therefore deserves some attention.

The telegraph network over which Orton presided used a modified version of the system introduced by Samuel Morse in 1844. Operators simply depressed a key that opened and closed a circuit, sending either dots or dashes of the Morse code, depending on the length of time they depressed the key. At the opposite end of the line, another operator took the message by listening to the clicks of a sounder and transcribing the message by hand. The Morse system dominated American telegraphy, and Orton believed it did so for a very simple reason. The Morse system transmitted the typical short business messages that dominated the long-distance telegraph market more rapidly than systems employing more complex mechanisms.[34] In contrast, European telegraphs transmitted a much higher percentage of government and social messages. Thus, in Britain the Wheatstone ink-recording telegraph was used extensively because there were greater percentages of government and press dispatches sent over the telegraph. However, during the 1870s, the British did adopt the Morse key and sounder system for most commercial messages. In France, government business dominated telegraph traffic, and improved versions of the printing telegraph first developed in the United States by David Hughes were used extensively on the main lines. Nonetheless, the Morse system was used for other types of messages in both countries.[35]

Because of the orientation of the American market toward business, Orton was concerned with the values of those users. He believed that the advantage of the telegraph to the businessman lay not in the savings in time it offered over mail, but in "its practical annihilation of time."[36] Any apparatus that caused unnecessary delays in the preparation or delivery of dispatches was no improvement. In Orton's view, the transmission of telegraph messages called for a very different kind of invention than that demanded by the normal processes of mass production. He argued that

> In all probability the telegraph will never run itself. Human intervention will always be necessary to some extent. The errors which, it seems to me, dreamers upon these subjects fall into, result from the attempt to treat ideas, and the intangible processes of their transmission

to a distant point, as physical things to be disposed of in bulk by the application of mechanism and power.[37]

Orton's belief in the central role of skilled human labor in telegraphy caused him to reject the principal system competing with the Morse.[38] The automatic telegraph systems introduced during the 1870s and 1880s were based on the original 1846 design of Alexander Bain. Promoters believed that automatic telegraphy's greater transmission speeds, combined with its ability to use unskilled labor to operate the machinery, gave it important advantages over the Morse. Proponents also argued that mechanically recorded messages were more accurate than those recorded by human operators.[39]

Although competitors began using automatic telegraphs, Orton remained unalterably opposed to them for several reasons he frequently articulated. First, and most important, the use of automatic telegraphs seemed to violate his dictate that speed of dispatch was the key to successful competition for telegraph traffic. The great potential advantage of the automatic—the actual transmission speed—also required economies of scale; only long messages, such as news stories or several short business messages transmitted together, enabled the greater speed of transmission to compensate for the delay in preparation of the perforated tapes and transcription of the messages. Orton objected to delaying business messages until a sufficient number had accumulated to send together and argued that the shorter period of time needed to prepare and transcribe an ordinary, individual message made the Morse system more advantageous for this type of business.[40] Second, he believed that any system of telegraphy needed skilled labor to ensure accurate transmission of messages. In Orton's view, automatic telegraphy required skill during both the perforation and translation of the messages, although proponents believed these tasks necessitated only unskilled labor. Third, contrary to the expectations of the system's promoters, the mechanical devices remained sufficiently complex to require skilled operators, thus obviating the advantage of cheap labor. Finally, they required a larger amount of more expensive apparatus. Orton tested the automatic systems offered to the company, but continued to believe that they would never succeed in replacing the Morse.[41]

The problems that automatic telegraphs encountered in actual practice bolstered Orton's opposition to the system. Among the most significant was the phenomenon called "tailing," which resulted when induced currents prolonged the signal, thus elongating the dots and dashes and making the recorded message unintelligible. Although later improvements largely overcame this problem, it never entirely disappeared. Another major difficulty was presented by the necessity of designing a fast, reliable perforator simple enough to be used by the unskilled young girls whom proponents considered

Figure 5.3. The mechanical perforator used in Thomas Edison's automatic telegraph system was cumbersome. The *Telegrapher* noted that, while the machine was "pretty" and a good male operator could prepare thirty-five words a minute, it suffered frequent mechanical breakdowns. The machines were also difficult to operate for the unskilled young women that advocates hoped they could use to lower labor costs. (*Source:* Cat. 558, ENHS.)

essential to making automatic telegraphy cost effective. Other practical requirements for successful competition with Morse telegraphy included repeaters to allow for long-distance transmission and some method of dropping messages at way offices.

Because Orton strongly believed that prompt, reliable, accurate service was best obtained from skilled human operators working on the Morse system, he rejected automatic telegraphy when introducing other new systems on Western Union lines. These systems, known as multiple telegraphs, increased the number of messages that could be sent over a single wire at the same time using the Morse key and sounder. They therefore helped save on capital costs by reducing the number of new wires the company needed to build in order to transmit the same number of messages. At the same time, they supplemented rather than replaced the Morse instruments, thus preserving the company's investments in equipment and in trained operators.

The advantages offered by multiple telegraphs became apparent to Orton when they were demonstrated by inventor Joseph B. Stearns. In 1867, Stearns, then president of the Franklin Telegraph Company, faced the problem of insufficient wires for the increasing volume of the company's business.[42] Lacking the necessary funds to undertake a construction program Stearns consulted general superintendent James G. Smith and decided to try duplexing the wires—that is, simultaneously sending two messages over a single wire in opposite directions. Stearns, whose knowledge of electrical science and telegraphy developed during his years as an operator and manager of the Farmer-Channing fire-alarm telegraph, combined elements of several existing designs into a device that increased the capacity of the Franklin wires by almost 50 percent.[43] However, Franklin Telegraph's directors failed to take advantage of the potential competitive advantages afforded by the duplex, refusing either to buy it from Stearns or to pay development costs. As a result, Stearns resigned from the company in 1871 and convinced Orton to allow him to conduct tests on Western Union's lines.

By the time Stearns approached Orton, the Western Union president had begun to appreciate the competitive advantages to be gained by controlling key patents and inventions, as a result not only of Western Union's takeover of the Gold and Stock Telegraph Company, but also its experience with the Page patent. In 1868, the same year that Stearns received his first duplex patent, Congress had granted a special patent for the induction coil to Charles Grafton Page, in order to restore the American claim to the induction coil after Heinrich Rumkorff was awarded the Volta Prize for the invention of this device. The issuance of this patent, tinged as it was with political overtones, became a serious matter to American telegraph companies when attorneys for Page's widow claimed that it covered the use of circuit breakers worked by electromagnets, an essential feature of all American telegraph systems. After initially rejecting Priscilla Page's demand that Western Union pay five hundred thousand dollars for rights to the patent, Orton and other Western Union officials consulted with a number of lawyers and technical experts, including Morse himself, all of whom advised that the patent was valid. Fearful that control of such an overarching patent by a competitor could prove damaging to the company, Orton agreed to purchase it. In 1871, Western Union paid twenty-five thousand dollars for a half interest and stipulated that it would actively defend the patent against infringement. The value of the patent was never fully established in Western Union's principal market because the company forced competitors into mergers before the conclusion of the suits. The Page patent was, however, upheld in cases related to urban telegraph services and companies in this market were forced to pay royalties to continue using instruments covered by the patent. Assuring such payments had been the principle reason that Priscilla Page and her attorneys stipulated that Western Union pay for and actively prosecute litigation for infringement of the patent. The company subsequently spent large sums of money for this purpose and Priscilla Page realized some seventy thousand dollars from royalties during her life. While the costs of the patent to Western Union were high and its actual financial benefit uncertain, the patent's potential to control the telegraph industry was widely recognized. Orton and other Western Union officials undoubtedly understood what control of such a patent meant as a competitive tool.[44]

Western Union's simultaneous acquisition of the Gold and Stock Telegraph Company reinforced Orton's appreciation for gaining control of such patents. From his experience with Gold and Stock Orton learned the value of providing support for important inventors. His support of Phelps's inventive efforts helped to enable Western Union to threaten Gold and Stock and force the merger. Furthermore, Gold and Stock's own competitive position and its ability to bargain with Western Union resulted from the strong patent position it had gained through support of inventors such as Thomas Edison.

One other factor may have encouraged Orton to allow Stearns to conduct experiments of his system on Western Union lines. The creation of Western Union's near-monopoly position after the Civil War stirred new fears about the corrupting power of large private economic interests. In 1869 Congress held the first of what would become semi-regular hearings on the question of whether a "telegraph monopoly" had arisen which was injurious to the nation and should be replaced by a government postal telegraph as was becoming common in Europe.[45] Although the question of innovation was a minor point raised in these hearings, which primarily focused on issues of rates and service, it was one that Western Union officials had to answer, especially as the company's leading critics were often supporters of new telegraph systems they claimed would reduce prices and provide better service. In 1869 Gardiner Hubbard, who drafted a bill calling for the creation of a new national telegraph company that would serve as a contractor to the post office, pointed specifically to the Franklin Telegraph's use of the Stearns duplex as an important invention that would increase efficiency and reduce rates. It was in response to Hubbard that Orton made his claim that it was in the company's best interest to adopt inventions by which the "despatch of business . . . can be materially increased."[46] Although Orton was not responding directly to critics when he decided to test the Stearns duplex, he could hardly ignore the potential public relations benefits, especially as his own experience was leading him to believe that the control of key patents and inventions could provide other competitive advantages as well.

Stearns began his new experiments on Western Union lines in the fall of 1871 with the assistance of company electrician George Prescott. Within six months he had radically improved his system by adding a condenser to

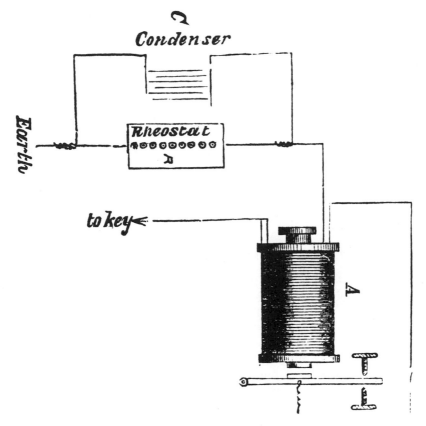

Figure 5.4. Joseph B. Stearns's successful duplex design (U.S. Pat. 126,847) used a condenser in the artificial line to store an electric charge sufficient to match the electrostatic capacity of the line and counteract the "static return charge" (known as a "kick"). This prevented the receiving relay from responding to outgoing signals.

overcome the effects of static discharge.[47] This improved duplex had an immediate and dramatic impact on Western Union's lines. Because the duplex nearly doubled the capacity of a wire, company officials found that they needed fewer new wires, saving construction costs. Fewer wires also meant substantial maintenance savings, as lines had to be reconstructed on an average of every twelve years. Even more important, in Orton's opinion, was the ability of the apparatus to increase the capacity of a line temporarily when wires failed or when particularly heavy traffic taxed the capacity of a line being worked in one direction only. Duplexing also allowed the company to better serve tourist resorts, which needed extensive communications facilities only during certain seasons. Orton came to view the Stearns duplex as the most important improvement in telegraphy since Morse's original invention. In a letter to its inventor, Orton stated that he considered it so valuable that he would not sell it for a million dollars. Thus, he moved quickly to acquire all the patent rights to Stearns's duplex system and to place it into service on the company's lines.[48]

The successful outcome of Western Union's support for Stearns's experiments and his growing awareness of the importance of patents in preventing use of a technology by competitors encouraged Orton to pursue a more aggressive strategy in regard to invention. He moved to prevent further use of the Stearns duplex by the Franklin Company and also engaged the services of another inventor to work on duplex telegraphy. Thomas Edison had come to Orton's notice through his work for Gold and Stock on printing telegraphs, which had given him a reputation for devising ingenious variations on the existing technology. After Edison approached Orton in the fall of 1872, claiming he could achieve comparable success with duplex telegraphs, the Western Union president wrote Stearns, saying

> I became apprehensive that processes for working Duplex would be devised which would successfully evade your patents, and also that your [Stearns's] attorneys had not done their work in shutting out competitors, as might have been done. I therefore, sent for Edison, and after several conferences and much discussion with him, I employed him to invent as many processes as possible for doing all or any part of the work covered by your patents. The object was to anticipate other inventors in new modes and also to patent as many combinations as possible.[49]

Edison put it more simply, claiming he was hired to invent duplexes "as an insurance against other parties using them—other lines."[50]

Edison's work on multiplex telegraphy reinforced Orton's interest in new inventions and led to an increase of company funding to inventors for specific projects. While working on duplex telegraphy with assistance from Prescott and the electrician's office, Edison developed an unexpected bonus for Western Union when he invented a practical quadruplex telegraph, by which two messages could be sent simultaneously in each direction. But Western Union failed to secure a formal agreement with Edison for control of any inventions he had developed with Western Union support, and this led to a protracted court battle with Jay Gould's Atlantic and Pacific Telegraph Company for control of the quadruplex. As a result, Orton also gained a new appreciation for the necessity of more formal relationships with inventors working for the company. Subsequently, when Edison had a falling-out with the managers of Atlantic and Pacific, Orton again acquired the inventor's services to make further improvements in multiple telegraphy and to prevent him from working for the opposition. This time, however, Orton drew up a formal contract

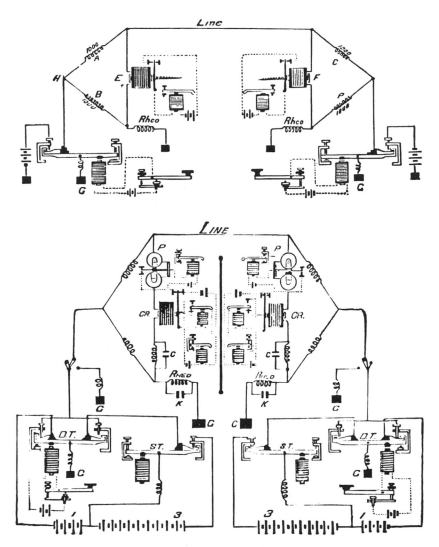

Figures 5.5 and 5.6. Francis Jones, later chief electrician of the Postal Telegraph Company, presented one of the earliest descriptions of Edison's quadruplex in a paper before the American Electrical Society in Chicago. These drawings of the circuit diagrams for Stearns's bridge duplex and Edison's quadruplex appeared side by side in the society's journal. (*Source: American Electrical Society Journal* 1 [1875]: 20–21.)

Figure 5.7. Elisha Gray later claimed that this mechanical transmitter designed for his harmonic telegraph produced a sound similar to the "human voice when in great distress" and suggested to him a means for transmitting speech. The pulley powered by a steam engine revolved the shaft on which sat two eccentric cams with projections on them that caused two small levers to vibrate upon their contact points and send harmonic frequencies out over the line. By means of adjusting springs he could change the pressure of the levers on the contact points and alter the frequency of their vibration, thus producing different sounds. (*Source:* Gray, *Experimental Researches,* 53.)

that called for Edison to develop a telegraph system that could send more than four messages at a time, in return for experimental expenses of two hundred dollars per week, as well as other assistance from the company.[51]

Thomas Edison was not the only beneficiary of Orton's farsighted strategy. Western Union also provided assistance to Elisha Gray, another inventor whose printing telegraph patents were controlled by Gold and Stock. Both Edison and Gray were working on acoustic telegraph systems designed to transmit multiple messages by means of different harmonic frequencies using tuning forks or reeds. Western Union first became interested in acoustic systems in May 1874, when Elisha Gray demonstrated his system of harmonic telegraphy to company officials. By 1874, Gray was well known to company officials for his printing telegraph inventions and for his work as electrician of the Western Electric Manufacturing Company of Chicago, which was partly owned by Western Union. Nothing immediately came of the demonstration and Gray went to Europe to exhibit it, but after his return he and the company renewed their discussions. The company apparently agreed to give him a five-thousand-dollar loan, using his stock (approximately one-third of the total interest) in the Harmonic Telegraph Company as collateral. Western Union also provided Gray with the facilities of Phelps's shop.[52]

Western Union's support for acoustic telegraph experiments paid off when Alexander Graham Bell, who was working independently on acoustic telegraphy, announced in May 1876 that his instrument could be made to transmit speech. Orton initially rejected Bell's offer to sell the new invention to Western Union, believing that it was technically inadequate for long-distance transmission. Edison and Gray therefore continued to focus their work on acoustic telegraphy. When Bell and his backers began using the telephone to compete with intraurban telegraph services, such as those supplied by Gold and Stock, thus threatening Western Union's own interests in this field, Orton responded by acquiring Gray's inventions suited to telephony, as well as encouraging Edison's work on the new technology. Edison's experiments resulted in a new contract with Western Union, which moved the company toward a potentially significant innovation in corporate support for invention.[53]

Prompted by the unexpected development of both the quadruplex and the telephone, Orton decided to support an ongoing research program rather than to pay an independent inventor for specific improvements in the company's apparatus or for work on a new device. Using money from his contract for work on acoustic telegraphy, along with royalties from his earlier telegraph work, Edison had built his Menlo Park, New Jersey, laboratory in early 1876. At the inventor's initiative, Western Union entered into a new agreement with Edison in March 1877 which provided general support for the laboratory's investigations related to telecommunications. During the negotiations over this contract, Edison had pointed out the advantages the laboratory offered for further enhancing his proven abilities to turn out new inventions with startling regularity. Under the new arrangement, Western Union provided one hundred dollars per week for the "payment of laboratory expenses incurred in perfecting inventions applicable to land lines of telegraph or cables within the United States."[54]

While Edison's laboratory was clearly perceived by Orton as an extension of the inventor himself, Western Union's willingness to support the laboratory was a step toward more direct corporate control of the inventive process. At the same time, Orton continued to provide significant assistance to other inventors in the Western Union fold, particularly George Phelps, superintendent of the company's factory, and Gerritt Smith, assistant to George Prescott.[55] Orton's sudden death in early 1878 ended his pursuit of greater control over the conduct of research important to Western Union. Although Orton might not have continued to provide such high levels of support,

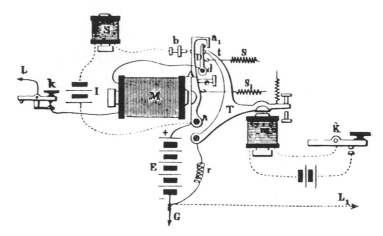

Figure 5.8. William Orton supported the work of Western Union assistant electrician Gerritt Smith, who produced a number of improvements in duplex and quadruplex telegraphs. Like other inventors working on multiple telegraphs, Smith focused on electrical designs, but he too produced mechanical solutions such as this electromechanical duplex in which an escapement mechanism T, operated by the electromagnet m prevented the armature A of the transmitting station's receiving relay M from responding to the outgoing current. In designing this device Smith hoped to obviate the need to use artificial lines or other forms of electrical compensation. (*Source:* Davis and Roe, *Handbook of Electrical Diagrams,* plate xxii.)

especially as conditions within the industry changed, his policies suggest he would have continued to pursue a broad program.

When William Orton developed his aggressive strategy of supporting new invention during the 1870s, telegraphy was the preeminent field of electrical technology on which inventors focused their efforts. By the 1880s, the forefront of electrical technology began to shift to newer fields of electric lighting and telephony. In both industries inventors originally connected with telegraphy played a key role and, in fact, Western Union's support of Edison and his laboratory can be seen as an important factor in the origins of the electric light industry. Nonetheless, the very success of Orton's strategy in helping Western Union secure industrial dominance enabled his successors to employ a more traditional approach to invention whereby the company acted largely as a passive consumer of rather than an active partner with the independent inventor. Yet, even under this traditional approach, the inventions and patents of the independent inventor were now clearly tied to corporate strategies designed to ensure long-term security in a competitive marketplace.

The Decline of Innovation

William Orton's successor at Western Union, Norvin Green, held similar views concerning the business of telegraphy, but he evolved a very different strategy with regard to invention. Green, who had been vice-president of Western Union at the time of Orton's death, believed that technical advantages derived from the duplex and quadruplex had helped Western Union secure its dominance of the telegraph industry. He argued that the improved service resulting from these inventions and other improvements precluded competitors from satisfying "the requirements of the business public, which, by the prompt service we have been able to do for the last three or four years, have been educated up to such exactions that they consider a delay of half an hour sufficient cause of action in claims for damages."[56] Under Green's direction, the Stearns patent became the basis of Western Union's patent strategy and was vigorously defended under the assumption that all subsequent developments in multiple telegraphy, including the quadruplex, were "obliged to use its condenser and other devices in modified form."[57] The threat of such suits caused competitors to seek noninterfering inventions in multiple telegraphy or in other telegraph technologies not controlled by Western Union.

Even before Orton's death, Green was using Western Union's strong patent position in telephony in his negotiations with the American Bell Telephone Company.[58] He also sought to gain greater security in Western Union's own market of long-distance message transmission. Both Green and Orton were willing to compromise in the competition over the telephone, because they viewed control of inventions in that field as a secondary goal to securing the company's dominance in its principal market. After taking over as president, Green eventually agreed to trade the company's telephone patents for a percentage of Bell Telephone's profits and, most important, a monopoly over intercity telegraph messages transmitted through intracity telephone exchanges. Furthermore, American Bell agreed not to compete for long-distance messages during the life of the patents.[59] As Leonard Reich has pointed out, this trading function is one of the key elements of modern research and development strategy.[60]

Although Green recognized the importance of controlling inventions and patents that might prove of value to the company, he appears to have been less inclined than Orton to support inventive activity directly. This decline of support stemmed in part from the perception of Western Union as a telegraph rather than a communications company. Thus, Green failed to perceive the long-term development of telephony as a threat and did not continue Western Union's support for telephone or multiple telegraph inventions. Even though Edison's 1877 contract with Western Union was still in force, American Bell later acquired rights to future telephone inventions made under it.

While instructing the electrician's office to monitor the work of outside inventors in order to protect the company from "expensive mistakes of either infringing a patent unnecessarily, or neglecting to secure such as promise to be of value to our business," Green virtually ended direct support for outside inventors.[61] The company's experimental shop was also moved from the headquarters building to the company's factory under the direction of George Phelps who became the company's principal inventor. With the electrician's office coming to play a less direct role in invention, Gerritt Smith, who had made many key contributions to the development of duplex and quadruplex telegraphs resigned his position as assistant electrician to join the rival American Rapid Telegraph Company.[62] Two notable exceptions to this policy during Green's tenure were the assistance provided to Stephen Field and David Brooks. Support for Field's work on a system of operating telegraph lines using dynamos in place of batteries and adapting it to run the quadruplex grew out of interest in engineering efficiency, while Brooks's development of an underground system of telegraph lines was supported in response to political agitation calling for the removal of overhead electrical wires.[63]

Green's more passive approach to new inventions resulted from a growing belief that long-distance telegraph technology was mature. This belief was created in part by the failure of inventors to develop an alternative system that could give an important technological advantage to either Western Union or its competitors. Western Union's large investment in both equipment and training in the existing Morse system also influenced Green's thinking. He was willing to support inventive work that supplemented rather than replaced this system, and he encouraged the engineering staff to improve its operating efficiencies.[64] However, he believed that the newer electrical fields of telephony and electric lighting offered greater opportunities for significant improvements than telegraph technology. Thus, Green responded to an inquiry from G. Kyle concerning the future prospects of electrical technology by recommending that Kyle enter one of the newer electrical fields rather than telegraphy. For the same reason Green supported Edison's experiments on electric lighting as first president of the Edison Electric Light Company at the same time that, as Western Union president, he questioned that company's continued support for the inventor.[65]

Green may also have felt that Western Union's extensive system of wires and offices, acquired through its railroad contracts and by its takeover of competitors, gave the company tremendous competitive advantages. In order to compete with this system, rivals would have to make large investments in lines as Western Union's control over duplex and quadruplex telegraphy

Figures 5.9 and 5.10. Political pressure caused by the increasing web of wires for a growing number of telegraph, telephone, and electric-lighting systems in city centers forced Western Union president Norvin Green to support David Brooks's development of an underground cable system. In Brooks's system, flexible cables consisting of copper wires insulated with cotton or flax and wrapped in jute braiding were distributed throughout the city in iron pipes filled with paraffin oil. (*Source:* Reid, *Telegraph in America* [1886], 811; Prescott, *Electricity and the Electric Telegraph* [1885]: 1076–77.)

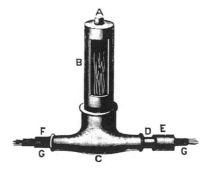

would prevent them from using these systems to reduce the number of wires they needed to expand their networks. Some competing companies responded by attempting to develop noninfringing duplex and quadruplex patents. But more significant technical challenges arose from rivals seeking to develop new systems, primarily automatic telegraphs, that might radically alter the technological and competitive structure of the industry.

The failure of companies attempting to compete with Western Union by using automatic telegraph systems demonstrates the difficulties facing competitors who pursued a technological strategy. Two major attempts were made to compete using automatic systems. The Automatic Telegraph Company made the first attempt in the early 1870s, followed by the American Rapid Telegraph Company in the early 1880s. Both companies sought to match Western Union's national system and service by building up lines through alliances with other telegraph companies. In both cases, they were taken over by larger conglomerates who subsequently abandoned automatic telegraphy because it did not drastically undercut Western Union's prices or improve on its service.[66] Although high-speed automatic telegraphy was plagued by many technical problems, its promoters failed primarily because they did not recognize that it was ill-suited to compete comprehensively with Western Union. Automatic telegraphs possessed certain advantages in the small part of the telegraph market that valued low-cost transmission of long messages over immediate transmission and reception; a market made up primarily of press reports and night letters.[67] Western Union itself adopted the low-speed, ink-recording Wheatstone automatic in 1882 for press service and night messages in order to free lines for business traffic.[68] American Rapid had initial success when it carved out a niche in the press service and night message markets, but its managers, as did other supporters of automatic telegraphy, failed because they did not recognize that businessmen had come to rely on immediate transmission and delivery of ordinary messages.[69] The higher transmission speeds possible with automatic telegraphs did not offset the time required to prepare messages for transmission and to transcribe them before delivery. For American Rapid, as for Automatic Telegraph before it, expanding to compete with Western Union for ordinary business messages meant selling out to companies less personally committed to automatic telegraphy and to a technological strategy of competition.

Western Union faced more significant competition from companies able to build large networks of lines using the Morse system. The many small companies seeking to compete with Western Union in the late 1860s and early 1870s were unable to make the necessary investments in line building, but in the mid-1870s, and again in 1880, Jay Gould used his railroad connections and large financial resources to build large competing networks. Western Union had created its national network in large part by its use of railroad rights-of-way for its lines. Its exclusive contracts also prevented rivals from using these same rights-of-way. Unfortunately, the company had not secured telegraph contracts and rights-of-way from Union Pacific or Central Pacific. These instead went to the small Atlantic and Pacific Telegraph Company. Gould's interests in these railroads, his rivalry with fellow railroad baron Cornelius Vanderbilt, the major stockholder of Western Union, and the strategic importance of telegraphy to the railroads and to American business in general, prompted him in 1874 to gain control of Atlantic and Pacific Telegraph as the basis for a new telegraph empire. He soon acquired an Atlantic cable connection, took over Franklin Telegraph, and sought telegraph rights from other railroad lines. By 1875 he also controlled Automatic Telegraph, which had been negotiating for railroad rights in order to build up its own network of lines.[70]

Technological innovation played a more minor role in Gould's strategy. The automatic telegraph was used on the Atlantic and Pacific lines because when it worked well it enabled the company to handle a larger amount of traffic than it might have without such a system. The automatic, however, never proved sufficiently economical and reliable to succeed in long-term competition with the Morse.[71] In addition to the automatic system, Gould also acquired rights to Edison's quadruplex and hired Edison as company engineer. Although Gould apparently was initially interested in possible technical advantages to be gained by supporting Edison, his primary interest was in damaging Western Union rather than pursuing innovation.[72] When Gould and Eckert supported the inventive work of Edison's successor as company electrician, Georges D'Infreville, it was because he devised a duplex telegraph that allowed the company to use that system without infringing on Stearns's patents.[73]

Gould failed to defeat the industry giant in the marketplace because he could not expand his system rapidly enough to take sufficient business from Western Union, especially after Western Union reduced its rates to meet the lower rates of Atlantic and Pacific. When Gould agreed to a merger with Western Union in 1878, Atlantic and Pacific had only 36,044 miles of wire and 1,757 offices, compared with Western Union's 194,323 miles of wire and 7,500 offices. Gould's failure with Atlantic and Pacific nonetheless provided him with significant experience that he soon used in a more successful battle for control of the nation's telegraph industry.

In 1879, Gould established the American Union Telegraph Company, which one year later boasted some 50,000 miles of wire and 2,000 offices.[74] The tremendous growth of this new opposition company was made possible by the same strategy Western Union had used to extend its lines—following the railroads. Gould was aided in his efforts by the passage in 1879 of a Congressional bill allowing railroads to conduct commercial telegraph business and by court decisions making it possible for rival companies to receive permission to use railroad rights-of-way, even when Western Union had exclusive contracts.[75] Through negotiations with others in the railroad industry who were dissatisfied with Western Union, Gould quickly built a rival network. In establishing his lines Gould entered into a series of arrangements with other railroads, including the Baltimore and Ohio, whose president was on the board of American Union, with other telegraph companies, with a cable company, and with urban telegraph companies. In this way American Union soon acquired important points of interconnection with other telegraph markets.[76]

The key to Gould's challenge was his ability to use the railroads to build up the American Union rapidly and make it a credible competitor. Although new technology did not play a part in competition between American Union and Western Union, the latter did seek to use its strong patent position to force Gould to use less effective technology. American Union thus endeavored to support invention and acquire patents that would allow it to avoid infringing on the key patents owned by Western Union—the Page patent controlling the adjustment of armatures by electromagnets, as used in relays and sounders, and the Stearns patent controlling the use of a condenser in duplex and quadruplex telegraphs. American Union tested several duplexes but declined to purchase them for fear of infringement after consulting with attorneys and technical experts. The company did adopt a noninfringing but poorly working duplex invented by Joseph Fenn, one of the company's assistant electricians, and also used the D'Infreville duplex, whose ownership status was unclear. As a consequence, Western Union promptly sued American Union for infringement of both the D'Infreville and Page patents. Soon after the Page patent suit was filed, the company's electricians were asked to design a noninfringing device, and chief electrician Henry Van Hoevenbergh invented an instrument that he began testing on the company's lines.[77] Infringement suits proved to be more of a nuisance than a real hindrance to American Union, and the battle was fought largely on nontechnical grounds.

Within two years of its establishment, American Union was the most formidable telegraph rival yet established and took considerable business from Western Union, whose receipts were reduced by about $5,000 a day or some $2 million a year. In 1881 this represented about 14 percent of Western Union's total receipts. Gould took advantage of the resulting depression in Western Union stock prices to purchase large blocks of its stock. This led to negotiations with William Vanderbilt, who had taken over the family's railroad and telegraph empire in 1877 following the death of his father, Cornelius Vanderbilt. As a result of these negotiations, Gould gained control of Western Union in early 1881.[78] The failure of companies that relied primarily on technology, such as those using automatic telegraphy, and the success of Gould's approach led other rivals to focus on finding alternative ways to build up competing networks of lines.

Gould's takeover of Western Union had an indirect impact on the company's strategy toward invention. From its emergence as the dominant telegraph company in 1866, Western Union had faced antimonopoly sentiment. But after Gould's takeover, members of the financial community and the press, who had previously found Western Union's size advantageous in its ability to serve their needs, expressed concern that, under the direction of the hated Gould, it would exercise undue control of the nation's economy.[79] Green, a former politician, focused much of his energies during the remainder of his tenure as president on public relations and the questions being raised concerning public regulation of the industry. Although many of the same arguments first raised in the 1860s continued to be used against the telegraph monopoly, Green was in a stronger position with regard to the company's support for new technology. Not only had the company adopted the duplex and quadruplex, but it had played an important role in the development of the telephone without creating a monopoly over this new communications technology. Increasingly the criticism of Western Union's failure to adopt new technology focused on claims concerning the advantages of automatic telegraphy and centered on the question of whether mechanization was appropriate to telegraphy.[80]

While the question of new technology remained a minor one in regard to the issue of monopoly, a new area of public concern and debate forced Green to support invention. In this case, Green was attempting to head off public regulation by state and city governments striving to drive the growing profusion of telegraph, telephone, and electric-light wires underground. Thus, even as Green lobbied against attempts to force wires underground, he was investigating underground systems with the intention of blunting criticism by gradually placing the company's lines underground in the major cities. To that end, he supported the inventive work of David Brooks, whose system was adopted by the company in 1879.[81]

The antimonopoly sentiment directed at Gould's Western Union failed in its attempts to nationalize the telegraph, but it did influence competitive strategies during the 1880s. Baltimore and Ohio Railroad Company president Robert Garrett had long opposed Western Union's attempts to assert a monopoly over intercity telegraph messages and to control railroad telegraphy through its contracts with the various railroad companies. In 1882, he used his railroad's telegraph system as a basis for establishing the Baltimore and Ohio Telegraph Company as an alternative to the hated monopoly. The new telegraph company vigorously pursued a program of construction and, in 1884, entered into an alliance with two other competing companies, Bankers and Merchants and the Postal Telegraph Company. The Bankers and Merchants failed shortly thereafter, ending the alliance, and the Baltimore and Ohio subsequently pursued an independent competition.[82]

In order to compete effectively with Western Union, Baltimore and Ohio did require duplex and quadruplex telegraph technology. Thus, in 1884 they acquired the duplex and quadruplex patents of Benjamin Thompson and Charles Selden, who was made company superintendent. These patents had been offered several times to Western Union, which declined to purchase them at the price offered, believing that they were covered by the Stearns and Edison patents.[83] Western Union filed a patent infringement suit against Baltimore and Ohio, but its motion for a preliminary injunction was refused. The court, while holding that the claims of this reissue were valid, thought that the reissue itself might be contested as it

> must have been obtained upon the theory that the patent might be useful as a weapon of offense by means of a claims so comprehensive and elastic as to embrace within their scope all subsequent inventions which might be made in the same field of improvement . . . The reissue in suit was apparently designed to overreach these patents and subordinate them to the complainant's monopoly.[84]

In this ruling, the court recognized the basis of Western Union's strategy regarding invention and called it into question. In 1887, with the fate of the legal battle uncertain, and tired of costly competition, Garrett agreed to Western Union's takeover of Baltimore and Ohio Telegraph.[85] This left Western Union with only the Postal Telegraph and Cable Company as a significant competitor.

Postal and Cable Telegraph slowly built an extensive competing system with Western Union by takeovers of other small telegraph companies and by using its important cable system to build up its business.[86] Surprisingly, Western Union made no move to take over this lone competitor and, in the 1890s, entered into a pooling agreement with it.[87] Western Union apparently adopted this strategy of cooperation in reaction to the growing tide of antimonopoly sentiment in America. The concern about monopoly power expressed by the court in the Baltimore and Ohio case was echoed by various citizen groups and witnesses before state and federal legislative committees. These concerns focused on the railroads and other large corporations as well and resulted in the passage of the Sherman Antitrust Act of 1890.[88] The rise of this sentiment and the movement to limit the power of large corporations probably convinced Western Union officials that allowing a small competitor to have a share of the market was to its own benefit. The existence of this competitor also helped prevent any new competition in the telegraph industry, and both companies increasingly focused their technical efforts on improving the engineering efficiency of their systems.

Companies in the telegraph industry were among the first to seek the competitive advantages made possible by controlling invention. By the turn of the century, companies in many other industries sought to exercise similar control. Recognizing that patents were "the best and most effective means of controlling competition," a few firms such as Bell Telephone, General Electric, and the United Shoe Machinery Company began to employ corporate inventors to produce inventions that not only improved their products but also created patents to help protect their dominant position in the market.[89] While the patent system remained the principal means by which society sought to encourage invention, that system was increasingly being used to secure the industrial property of large, economically powerful corporations rather than the intellectual property of individual inventors. Changes in the postwar telegraph industry reflected a general transformation taking place in American society, as economic and political power began to shift from individuals to large-scale bureaucratic organizations.

6

From Shop Invention
to Industrial Research

A S THE telegraph became increasingly perceived as a mature technology, it began to acquire what historian of technology Thomas Hughes calls "technological momentum," that is, an inertia of motion that becomes difficult to alter.[1] Such momentum results from a combination of experience and investment in an existing technical system. In telegraphy, managers and engineers were committed to the Morse system by virtue of their practical training and experience; their technical expertise and personal career advancement were intimately tied to their origins as operators. They also acted to conserve the large investment their companies had made in training and equipment based on the Morse system. Because of this they believed that corporate interests were best served by focusing on the efficiency of existing systems, rather than on the invention of new ones.

As engineering tasks came to predominate in the 1880s, many of the independent inventor-entrepreneurs found their careers in telegraphy circumscribed. In order to retain their independence, they moved into other fields and acquired new support for their work. Those who remained in the corporation as inventor-engineers came to personify engineering values of standardization and efficiency.[2] The rise of engineering values within telegraphy was part of the larger development of professional engineering within American society, which was itself directly influenced by the growing importance of science-based sources of technological knowledge and training.

The growing importance of science to certain technical fields gave rise to a new organizational pattern for invention—the industrial research laboratory. Derived from European scientific research laboratories, industrial research laboratories were initially confined to large electrical and chemical companies. These laboratories differed from earlier existing organizations derived from traditional shop invention in their emphasis on the scientific backgrounds of researchers and in the extent to which scientific knowledge dominated the research process. Together the rise of engineering science and scientific industrial research hastened the decline of the mechanical tradition of shop invention which had predominated in American industry and had so influenced telegraphy.

Invention and Engineering

The growing importance of engineering solutions in the 1880s and 1890s should not overshadow the continuing similarity between inventors and engineers in telegraphy. Telegraph engineers derived technical knowledge from the same sources that the industry's inventors relied upon—practical experience at the operator's key or mechanic's bench and independent study in technical journals and books. Indeed, the engineering corps was largely made up of inventor-engineers, who replaced inventor-entrepreneurs as the principal developers of telegraph technology. Both emerged from telegraph shop culture with its origins in the tradition of American mechanical invention.

In some respects the growing emphasis on engineering resembled an earlier period of system engineering in the late 1850s and 1860s, when the growth of large telegraph systems first produced a need for improved components such as repeaters and insulators to increase operating efficiency. In that era, however, telegraph companies relied on the individual initiative of operators, managers, or manufacturers to produce technical improvements as independent inventors. By the 1880s, telegraph companies had established engineering and patents departments for supervising ongoing technical development. The men employed in these departments not only reviewed the work of outside inventors but undertook inventive work themselves. They were primarily engaged in solving engineering problems and developing standardized procedures for the technical operations of their companies.

Standardization of practice was only one element of engineering activity. Creative design was a key characteristic of the professional engineer, and in telegraphy such design often evolved from inventive activity.[3] The men who staffed engineering positions in the industry produced minor but important inventions designed to increase the efficiency and economy of the systems under their charge. With the expansion of staff technical positions in the 1870s and the employment of competent men to fill them, companies began turning to their electricians when faced with technical challenges from competitors.

Like earlier inventors, most technical personnel began their careers in the operating corps and were self-educated in electricity. Many engaged in invention before they assumed more responsible engineering positions and used their creativity to demonstrate their technical skills. Although Francis Jones, the chief electrician of Postal Telegraph, advanced farther than most, his career was in many ways typical of those who became members of corporate engineering staffs. Jones began his career in 1867 as an operator in St.

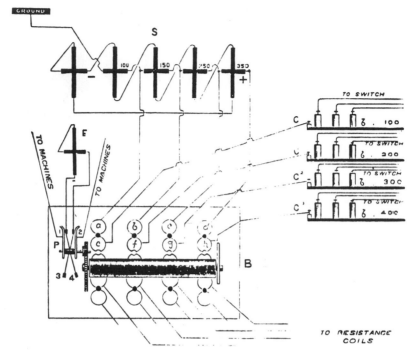

Figure 6.1. Francis Jones's 1880 paper before the American Electrical Society in Chicago presented one of the first technical discussion of Stephen Field's system for working telegraph lines by dynamos rather than batteries. (*Source: American Electrical Society Journal* 5 [1880]: 67.)

John, New Brunswick. After moving to Western Union's Chicago office in 1873, he continued his study of electricity and began experimenting with duplex and quadruplex telegraphs, devising improvements that were adopted by the company. Jones also became secretary of the recently founded American Electrical Society in 1875 and was rewarded for his demonstrated technical abilities with the newly created position of general circuit manager. Illness forced Jones to resign from Western Union, but he later worked as electrician for the rival Bankers and Merchants Telegraph Company. In that position he continued to invent, devising the only duplex telegraph that worked adequately on long lines and did not use Stearns's condenser design. After Bankers and Merchants failed, Jones became chief electrician of the Postal Telegraph Company. There he invented a system to run quadruplex telegraphs by dynamos which did not infringe Stephen Field's system used by Western Union.[4]

Most of the time telegraph companies wanted their engineering staff to make improvements to existing systems rather than design inventions that circumscribed existing patents. George Hamilton, who replaced George Prescott as Western Union's chief electrician in 1880, originally joined the electrician's staff in 1875 after serving as an assistant to noted inventor and electrician Moses Farmer. Although Hamilton did not take out any patents, he assisted Stephen Field's experiments with dynamos and subsequently supervised the introduction of the Field system. Hamilton was later responsible for important but unpatented improvements to the system that greatly increased its efficiency.[5] Hamilton's work and that of his engineering staff during the 1880s reflected the growing concern at Western Union with problems related to the efficiency of existing systems.

During the 1890s, Western Union's willingness to enter into pooling agreements with Postal Telegraph encouraged managers in both companies to focus their competitive strategies on improved service. Though this had been a major concern of Western Union's since the 1870s, its competitors usually sought to reduce their prices in order to attract business away from Western Union, even as they sought ways to compete with its national service. Postal Telegraph official Edward Nally testified in 1910 that the company had given the question of service competition "more study than anything else, because we believe it resolves itself into a question of service" and that customers were most concerned with quick delivery of messages. This concern led the company to seek greater efficiency in its technical operations and to develop a special rush service using call boxes and messengers. The president of Western Union, Robert Clowry, used the same claim of service competition to explain the company's refusal to use an automatic telegraph system developed by inventor Patrick Delany. His argument echoed those used by William Orton and Norvin Green in the 1870s. Efficient use of automatic telegraphs required high volume to gain the advantages of high-speed transmission. This meant allowing messages to accumulate, and he considered this unacceptable to the company's customers. Although both companies continued to examine the work of outside inventors, those like Patrick Delany who developed new telegraph systems found officials at Western Union or Postal Telegraph generally unwilling to adopt them.[6]

Delany's career reflected the problems faced in the mid-1880s by inventor-entrepreneurs who tried to achieve independence and financial success from telegraph inventions. He began his career in typical fashion as an operator. His technical skill led to a series of appointments to management positions, including that of superintendent of the Automatic Telegraph Company. Delany began inventing while working for Automatic Telegraph, but he gave up telegraphy to become a newspaper correspondent after the failure of that company. In 1880, however, he decided to seek his living through invention. During the next two decades, he developed both automatic and multiplex telegraph systems and a host of minor telegraph inventions. While

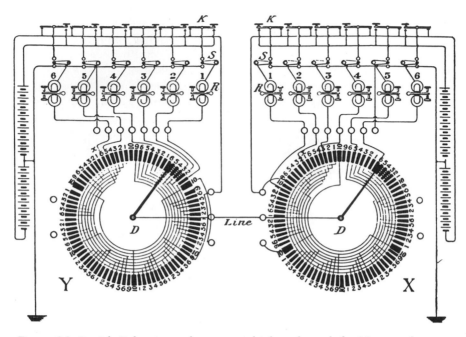

Figure 6.2. Patrick Delany's synchronous multiplex telegraph for Morse and printing telegraphs was adopted in Britain, but found little use in the United States. Its system of time-sharing, based in part on the phonic wheel design of Danish inventor Paul LaCour, was similar to that later employed in multiplex, page-printing telegraph systems. In this diagram the system is arranged for six operators whose instruments were connected to the line whenever the rotating pointer (known as a "trailer") came into contact with their segment of the wheel. With the trailer making about three revolutions per second, each operator would be connected with the line thirty-six times per second. (*Source:* Maver, *American Telegraphy,* 339.)

Delany found support for his new systems from independent investors and foreign governments, he found little support among the established American telegraph companies. His automatic system eventually became the basis of the Telepost Company's night-letter service. It was also tested over a British-owned Atlantic cable, but Western Union refused similar tests over its land line.[7] Though the British Post Office telegraph adopted his multiplex telegraph, its use in America was limited to the short-lived Standard Multiplex Telegraph Company, formed by a group of investors willing to speculate on Delany's efforts.[8] Western Union did, however, adopt Delany's design for automating the adjustment of relays at intermediate stations in bad weather.[9]

Telegraph companies were willing to invest in engineering inventions that increased the efficiency or improved the economy of their existing systems, but they were little interested in alternative telegraph systems. Independent inventors like Delany might produce engineering inventions, but they seldom moved into the corporation as engineers. Inventor-entrepreneurs who did become salaried employees for a company often found that their desire for independence conflicted with the goals of management, as occurred when Edison became electrician for Atlantic and Pacific. Jay Gould's decision to make Edison company electrician probably stemmed more from his desire to monopolize Edison's talents as an inventor, thus denying them to Western Union, than it did from his interest in Edison's services as an engineer. During the first six months of 1875, Edison did act in an engineering capacity in order to establish his automatic system on the A&P lines. He also offered suggestions for improving the company's operations. However, Edison's primary interest was in continuing to invent. Although he retained the title of company electrician for Atlantic and Pacific, Edison gave up active participation in its daily affairs because company president Thomas Eckert took a different view of the relative importance of Edison's two roles.[10] Edison was unwilling to act as engineer for the company if this meant that his inventive work would be circumscribed by corporate goals. Although Edison was forced to turn again to Western Union to support his laboratory independently of manufacturing, he did so on terms that provided relative freedom for his experimental work. Edison used his successful inventive work for Western Union to achieve a large degree of independence. By the end of the decade, as opportunities for significant inventive work in telegraphy declined, he used that independence to turn his attention to developing inventions for other electrical technologies.

Edison was among a small group of men who sustained careers as independent electrical inventor-entrepreneurs in the era of the large corporation. They did so by following Norvin Green's advice and turning their attention to problems in newer electrical fields such as electric light and power. Stephen Field, for example, moved from telegraph invention to work on developing electric railroad technology. Independent inventors such as Delany or Elisha Gray, who continued working on telegraphy, found that independent investors, European governments, and cable telegraph officials were more interested in their work than the managers of American telegraph companies.[11]

British telegraph official William Preece commented on the decline in support for nonengineering inventive work in the American telegraph industry, noting that "in 1877 I saw a good deal to learn, and picked up a great many wrinkles, and brought from America a good many processes. I go back there now in 1884, seven years afterwards, and I do not find one single advance made . . . while we in England, during those seven years, have progressed with great strides, in America . . . their scientific progress has not marched with their material progress, and invention has to a certain extent there ceased."[12] Yet Preece did adopt a new American invention on the telegraph lines of the British Post Office—Patrick Delany's synchronous multiplex. It was not telegraph invention per se that he found wanting, but rather the willingness of telegraph companies to adopt them. American telegraph officials clung to their prejudice for the Morse system as well as to their belief that the use of existing systems provided sufficient operating capacity.

Through the end of the century, the continuing dominance of the "broker" or short business message in telegraph traffic reinforced the bias of American telegraphers toward the Morse system. However, critics believed that by emphasizing the expeditious handling of broker messages, telegraph companies failed to attract more general business and social messages. Because other types of messages were longer, they were suitable for transmission by machine systems where high volume was more important than prompt handling. Because speed was a secondary consideration for customers sending such messages, the greater time necessary to prepare such messages for transmission and delivery would not obviate the advantages of higher transmission speeds.[13] Only when the market for long-distance communication changed, due to competition from long-distance telephony in the early twentieth century, did telegraph company officials begin to rethink their commitment to the Morse system and to respond more favorably to systems of rapid telegraphy.

Labor Relations and the Decline of Shop Culture

The commitment of telegraph leaders to Morse technology sprang from their common origins within telegraph shop culture. System managers and engineers began their careers in the ranks of the operating corps prior to the Civil War. They were immersed in the culture of the operators and committed to widely held views that the "knights of the key" were intellectually superior to most working men and women and to their fellow operators in Europe. At the end of the century, this attitude continued to inspire the expectation that the elite workforce of American operators would remain the principal source of managerial and engineering talent.[14] Telegraph companies continued to recruit managers and electricians from the ranks of operators and assumed that technical skill remained invested in operators, thus making the operating room a vital source of technical education within the industry.

Management's belief in the superiority of the Morse key and sounder system was also sustained by a number of other factors. Foremost was the belief that it provided the most efficient means for meeting the needs of business customers. Telegraph managers argued that only the Morse system allowed instantaneous transmission without preparation and did not require the backlog of messages to achieve economies of scale required by machine systems. This allowed messages to be sent almost immediately in the order in which they came to the office. Furthermore, they believed that the Morse system reduced errors by allowing skilled operators to make corrections during the course of transmission. The constant threat of lawsuits over transmission errors provided incentive to reduce such errors, but American telegraph officials believed that the Morse sounder and key system provided the best means to accomplish this.[15] The adoption of the Morse system by the British Post Office telegraph in the late 1870s combined with the failure of competing systems in the United States further confirmed American faith in the key and sounder.[16] Finally, management's desire to preserve the investment that their companies had made in the training of operators further reinforced their commitment to Morse technology. The only significant change in the system following the Civil War was the adoption of duplex and quadruplex forms of multiple telegraphy. These increased the number of messages that could be handled by Morse instruments on a single wire but did not radically alter the nature of the system itself.

Though the continuing commitment of telegraph managers to Morse technology helped to preserve the subculture of operators, labor conditions in the industry increasingly made the operating room less attractive and reduced its role as source of technical talent. Operators had begun losing status within telegraph shop culture during the 1870s. As the number of telegraph companies declined, operators faced diminishing opportunities for advancement. Declining opportunity also meant lower pay. A growing number of female telegraphers, encouraged by companies that sought to lower labor costs, particularly in low-traffic small town offices, further reduced the status of operators.[17] In 1878, railroad telegraph superintendent George Bliss took note of the changing status of telegraph operators and the reduced number of available positions, remarking that there was probably no trade in which the prospect for official advancement or for increased salaries was so small. As a consequence, technical expertise no longer provided the best means for taking advantage of those few opportunities that did open up. Bliss advised operators that the "strong disposition at the present time to become familiar with the scientific features of electricity" would pay few dividends "unless this is made a specialty." Instead, he urged operators to "confine one's self to the practical part of the business, which comes into constant use."[18] In contrast, Western Union's *Journal of the Telegraph* continued to express management's assumptions that it provided opportunity for those of "special aptitude" or "long service and practical ability." The *Journal* continued to urge scientific study as a means of advancement.[19]

The rise of telegraph-engineering departments during the 1870s contributed to the declining importance of technical proficiency in the career ad-

Figure 6.3. The Delany line adjuster electromechanically compensated for escapes and other disturbances on the line that weakened the signal and prevented local relays that were not properly adjusted from responding. Delany's instrument used the back-and-forth motion of pallet P on the armature lever to move wheels W and W', thus opening and closing the circuit between c and c' and causing the batteries at both ends of the line to open, thus clearing the line and allowing the local relays to respond. (*Source:* Maver, *American Telegraphy*, 152.)

vancement of operators. Formerly, operators had earned promotion and status by taking an interest in the mechanical operation of the system and developing skills and knowledge that went beyond mere manipulation of a telegraph key. As companies began to modify the engineering details of the system in order to rationalize their operations, the role of operators in day-to-day technical activity became increasingly circumscribed, especially after circuit managers were placed in charge of the testing and maintenance of lines. The use of dynamos to supply current on main lines beginning in the 1880s also greatly reduced the role of operators in line testing and the testing and care of batteries. By the end of the century, batteries were used primarily on shorter and less trafficked lines. The introduction of the Delany line adjustment system reduced the need for operators to make fine adjustments on their instruments. Telegraph companies sought to make such instrument adjustments as simple and routine as possible. Only on the quadruplex circuit, where delicate balancing of the line was necessary, did operators retain a large measure of control over the working of the instruments. Indeed, it appeared to many operators that service on the quadruplex circuits between New York and other major cities provided the best circumstance for promotion.[20] Engineering rationalization combined with declining opportunities for advancement to reduce the incentive for the independent study of electrical science and technology which had been a hallmark of the many operators who became inventors, engineers, and managers in the industry's first three decades.

The redefinition of operator skills and tasks was part of a pattern evident in other industries as managers emphasized increased dexterity rather than the range of skills and knowledge that had formerly marked the work of skilled machine operators. In his study of labor in the auto industry David Gartman argues that degradation of work was not always the intended result of technological changes.[21] This appears to be the case in the telegraph industry, where engineering improvements and the desire for higher transmission speeds caused the formerly wide range of operator skill and knowledge to be replaced by a narrow focus on greater dexterity. The failure of Western Union to develop automatic telegraphy, one of the few forms of technological innovation designed specifically to reduce the need for highly trained operators, suggests how other factors dominated management decisions in regard to labor skill. Company presidents Orton and Green both believed that the potentially higher transmission speeds and lower labor costs offered by automatic telegraphs did not offset the slower delivery of messages caused by the need to translate messages into machine-readable form before transmission and from Morse code into English afterward. They equated good service with skilled Morse operators who could instantly translate handwritten messages into Morse code and code into deliverable English-language messages. They further argued that this reduced transmission errors because skilled Morse operators could correct for incomplete or garbled messages caused by transmission problems. Nonetheless, this did not prevent the company from taking advantage of new technology that increased transmission speeds and reduced other operator tasks.

Telegraph operators themselves contributed to the redefinition of telegraph skills by diminishing the status attached to lost technical responsibilities in favor of greater proficiency at manipulating the key and higher transmission speeds. Patrick Delany, who was himself a former operator, noted that

> nearly all operators, good and bad, are vain of their abilities to send rapidly, and nearly all are ambitious to send faster than the operator at the receiving station can write it down. . . . Each seems individually impelled to "salt" the man at the other end of the line, if possible, and when he succeeds in making him "break," he mentally records a victory and goes at it again with renewed vigor. To outsiders this self-

imposed rapid pace may seem foolish, but to the knight of the key there is great glory in it.[22]

This practice was apparent to William Preece and Henry Fischer when they visited the United States in 1877. They attributed this pride in manipulating the key in part "to the manner in which Exceptional skill is commented upon and eulogized in the Telegraphic literature." However, they also recognized that "the prime cause for the desire to become noted for rapid working consists in the fact that exceptional ability is almost certain to be rewarded by an arbitrary increase of pay."[23]

The distinction operators attached to transmission speed led them to institute a significant technical change that had important long-term consequences for the operating corps. First-class operators, especially those who received press reports, introduced typewriters into telegraphy as they sought to improve their rate of reception. The positive reception given to typewritten messages led the telegraph companies to increase the salary of operators using them. As the typewriter came into greater use and increased reception speeds, it created a bottleneck at the transmitting end, which led to the introduction of the Martin mecograph, a semi-automatic transmitting key. Because it reduced the physical strain on operators through its sideways motion as well as by its ability to repeat dots, the mecograph increased transmission speeds. In the process, however, these changes placed even more pressures on operators by raising average rates of transmission. By the turn of the century, operator shop culture was fast disappearing as the operating room took on characteristics of the industrial factory.[24]

Operators found themselves unable to effectively counter declining wages, long hours, work speedups, and arbitrary hiring and firing policies. Their greatest effort came in 1883 when the newly organized Brotherhood of Telegraphers in affiliation with the Knights of Labor struck against Western Union. The formidable power of the industry leader soon broke both the strike and the resolve of operators, and there were only sporadic organizing activities and occasional and ineffective wildcat strikes during the remainder of the century. Operators did not successfully organize against Western Union until the turn of the century, but a major strike led by the Commercial Telegraphers Union in 1907 was no more unsuccessful than earlier strikes. However, changed conditions in the industry caused the 1907 strike to have major consequences for telegraph operators as companies actively began to replace skilled Morse operators with women operating new telegraph technology that employed keyboards similar to those found on typewriters operated by women in the office workplace.[25]

Around the turn of the century, several printing telegraph systems were designed that printed messages on page forms, rather than on strips of paper tape as had been the case with earlier printing telegraphs. The transmitters

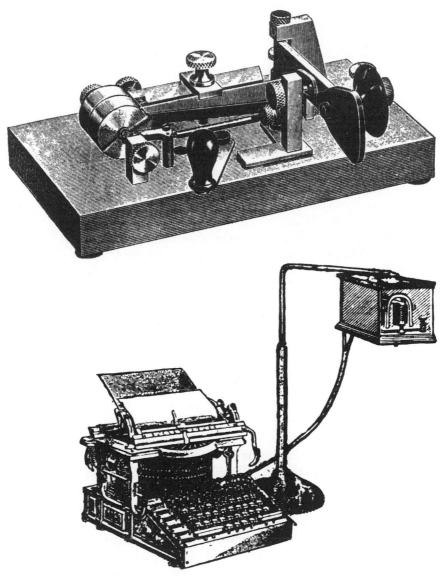

Figures 6.4 and 6.5. The Martin mecograph and the typewriter combined to greatly reduce the physical strains on operators, but they also placed new pressures on operators by raising average transmission speeds. The side-to-side movement of the mecograph, which also automatically transmitted dashes when the lever was placed on one side, was less taxing than the up-and-down movement of the traditional telegraph key. The typewriter allowed operators to keep up with higher speeds by reducing the fatigue of writing out messages by hand. (*Sources:* Maver and McNicol, "American Telegraph Engineering," 1319; Maver, *American Telegraphy,* 76.)

in these systems employed keyboards that required no knowledge of Morse code. Changing patterns in business use of the telegraph had encouraged telegraph companies to begin investigating the use of these systems on long-distance lines in place of the Morse system during the 1890s. Initial interest apparently grew out of a favorable response by customers to the growing use of typewriters by receiving operators. Indeed, business customers preferred the printed message. By the end of the decade, the development of printing telegraphs was also stimulated by the increasing use of long-distance telephone lines to transmit short business messages that formerly had been telegraphed. This "stimulated the desire for a formal method of intercommunication which is more rapid than the mails" among business customers and promoted the development of night-letter systems. Longer individual messages led to increased message density on telegraph lines.[26]

Western Union initially used the Buckingham-Barclay printing telegraph system developed by company patent attorney and inventor Charles Buckingham, and subsequently improved by company engineer John Barclay., While Western Union adopted the Buckingham-Barclay printer, Postal Telegraph adopted a system devised by American physicist Henry Rowland. Both companies also investigated the system of British inventor Donald Murray. All of these systems (except Rowland's, which used the keyboard directly as a transmitter) employed keyboard perforators to prepare a punched tape, similar to that used in the Wheatstone automatic telegraph, and printed messages on a page form, using a typewriter-like receiving instrument. Although the Rowland system was considered to be "very highly developed on the most modern and approved scientific principles," it was also considered much more complicated mechanically than the Buckingham, which proved better under conditions of actual use, particularly after Barclay improved its speed by using a modified electric typewriter as a simple, rapid printer. The Rowland system was designed to be multiplexed, but failed to perform adequately in this mode over long lines. This factor and its complicated design, which resulted in lower speeds, caused Postal Telegraph eventually to discontinue its use. The multiplex feature of the Rowland system, however, was to be an essential component of the teletypewriter form of printing telegraph developed by Western Electric and Western Union engineers before World War I, and which largely replaced the Morse key and sounder system following the war.[27]

While changes in the nature of telegraph traffic were encouraging the telegraph companies to rethink their commitment to the Morse system and to respond more favorably to systems of rapid machine telegraphy, the machines themselves allowed managers to redefine operator skills in a way that enabled them to alter fundamentally the makeup of their increasingly restive workforce. At the heart of this change was the new technology's keyboard.

Figure 6.6. Physicist Henry Rowland devised this synchronous multiplex printing telegraph system, which used no coded messages, transmitted directly from the keyboard shown here, and recorded messages on a page-printer. Using a time-sharing system similar to Delany's multiplex, four operators were each given the use of the wire for one-fourteenth of a second 3.5 times a second. (*Source:* Potts, "Rowland Telegraph System," 511.)

Although both male and female operators had been using typewriters since the mid-1880s, they remained ancillary to the skill of key and sounder and the operator's knowledge of Morse code. However, as Margery Davies has shown, elsewhere in the American workplace the typewriter was being defined as a feminine technology used primarily by the growing numbers of women who joined the office workforce of corporate America.[28] As the new telegraph technology made traditional operator skills unnecessary, its typewriter-like keyboard allowed managers to redefine telegraph operating as a female occupation akin to that of the office typist, thus ending the dominance of male Morse operators and their shop culture in the telegraph operating room.[29]

Shifting Patterns of Technical Knowledge

Even as shop culture declined among the operating corps, the continued dominance of the Morse system until after World War I reinforced the largely practical nature of American telegraph practice. The scientific principles underlying telegraphy did have importance for the advancement of the art, and inventors and engineers read scientific works on electricity as well as publications on telegraphy. However, telegraph technical literature focused on the design of machines, not the development of electrical science. Furthermore, engineering personnel continued to be recruited from the ranks of operators, who remained rooted in the shop tradition, rather than from the new electrical engineering schools. The decline of the nineteenth-century shop culture, and the use of the new information and tools being created by a cadre of scientifically trained researchers, left telegraph engineers ill-prepared to face challenges from the new telecommunications technologies of telephony and radio in the twentieth century.

The practical character of American telegraphy was fostered by the nature of its technical problems. The contrasting scientific influences on American and British practice were explained by Preece in 1877 when he noted that the

> absence of Submarine Cables, underground wires and complicated apparatus, and the climate, require less attention to the more abstruse laws of Electricity than has been the case in England. At home the intricate laws of induction have not only called forth the closest attention and study of the Telegraph Engineer, but the operations and researches of the Engineer have materially advanced our knowledge of the Science itself. Many new laws and striking facts have emanated from the practical telegraphist. Hence the English Telegraph Engineer has become essentially a scientific man.[30]

In British telegraphy, the problem of induction, particularly in regard to submarine cables, gave rise to a mathematical theory of telegraph transmission. William Thomson, professor of natural philosophy at Glasgow University and one of the premier physicists of the day as well as a telegraph inventor and engineer, was the leading figure in the development of this theory. Preece himself later developed an equation known as the KR law to determine the limits of telegraph and telephone transmission. Preece attempted to develop a rigorous engineering science of telegraphy, though he based his law on empirical analysis. Such empirical analyses ultimately proved inadequate, and a new generation of university-educated electrical engineers turned instead to knowledge and mathematical techniques from theoretical physics to develop an engineering science.[31]

In the United States, the problem of induction was most prevalent in high-speed, chemical-receiving automatic telegraph systems, although some difficulties also arose in the operation of duplex and quadruplex telegraphy. The bias of American telegraph officials against automatic systems meant that only a small group of inventors worked on the technical problems they presented. These inventors generally proceeded by empirical means, although they might develop theories to guide their experiments. Edison, for example, explained the effects of induction in one of his notebooks in the following manner:

> The transmission of waves of electricity are instantaneous no matter what the length of the cable. The retardation noticed by electricians in cables is ~~due to the~~ not properly retardation but the leyden charge sending its current against the charging current, the same as an electro-magnet placed in an electric current. The first part impulse will be weakened by the counter charge against the magnetizing current.[32]

Edison used this theory to develop a line-balancing method in which the receiving instrument was placed at a neutral point between an artificial line at the receiving end and the transmission line, both having equal resistances. Edison apparently determined his neutral point empirically by adjusting a rheostat in the artificial line to find the center of resistance. His patent on this method made no mention of his underlying theory and he made no attempt to develop an equation to help him determine the placement of the artificial line.[33]

Unlike their British counterparts, those American telegraph inventors and engineers who published articles and books or presented papers at professional meetings made no effort to develop general laws of transmission or to generate engineering formulas to use in circuit design. Prior to 1911, the only paper on telegraph transmission read by a telegraph inventor or engineer at a meeting of the American Institute of Electrical Engineers consisted entirely of descriptions of solutions appearing in patents.[34] Indeed, the unusual character of Frank Fowle's 1911 paper prompted one of the commentators to remark that it was

> of particular interest, as it undertakes to present and to treat in a scientific manner a number of telegraph problems in the same way which has been so successful in dealing with problems concerning telephone work, power transmissions and electric light work. But few papers have been presented recently in this country which have dealt with telegraph problems as though they were controlled by the same physical laws as are the other branches of the electrical art.[35]

The mathematical formulas presented in a major American work on telegraphy published in 1913 consisted entirely of simple arithmetic and elementary algebra. *Telegraph Engineering,* the first significant telegraph work to require an understanding of calculus in order to follow some of its mathematical demonstrations, was written two years later by Erich Hausmann, a professor of physics and electrical engineering at the Polytechnic Institute of Brooklyn.[36]

The contrasting approaches to the electrical analysis of duplex and quadruplex systems undertaken by William Maver in the 1880s and by Stanley

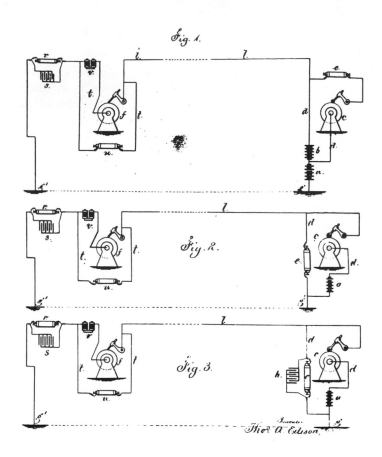

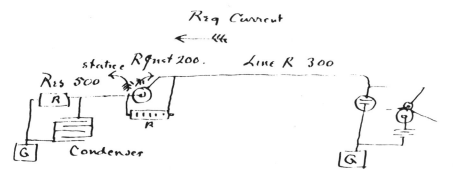

Figures 6.7 and 6.8. Edison's notebook entry describing his "system of Cables working by Centers of resistances and static accumulation" noted that it worked perfectly between New York and Washington at 1,600 words per minute. The patent (U.S. Pat. 147,311) showed several alternative designs. (*Source: TAEM* 6:46.)

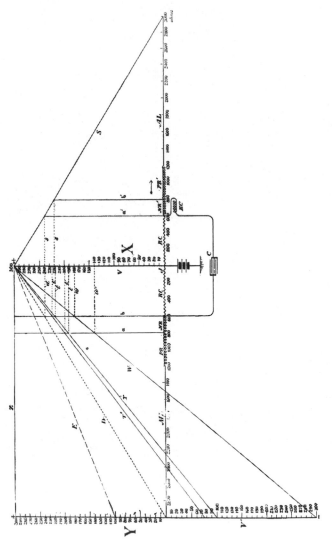

Figure 6.9. In his book *American Telegraphy*, William Maver used this graphical representation of the changing potential of condenser plates caused by battery reversals in Gerritt Smith's quadruplex design. (*Source*: Maver, *American Telegraphy*, 205.)

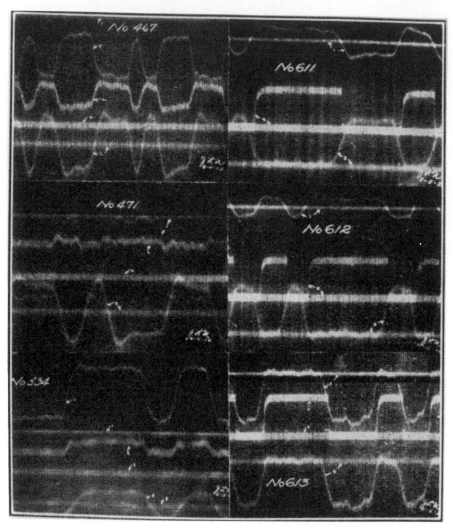

Figures 6.10, 6.11, and 6.12. Stanley Rhoads used an oscillogram to analyze the line current in quadruplex telegraphs under differing conditions and devised formulae for calculating line current values. Like Maver, however, he presented his findings in an easily understood visual form. (*Source:* Rhoads, "Railroad Telegraph and Telephone Engineering," 340, 355–56.)

1. For the condition of keys closed:

$$I = \frac{E}{R + \sqrt{\dfrac{r}{g}} \ \tanh \dfrac{L}{2} \ \sqrt{r\,g}}$$

2. For the condition of distant key open:

$$I = \frac{E}{R + \sqrt{\dfrac{r}{g}} \ \coth L \ \sqrt{r\,g}}$$

E is voltage at one terminal, R is terminal resistance of one terminal, r is average line resistance including relays, g is the leakage conductance, mhos per mile, and L is the length of line in miles.

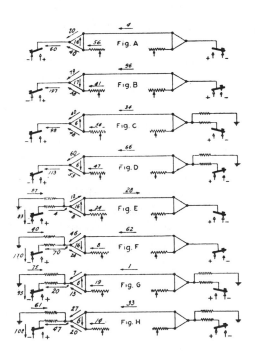

Rhoads in 1921 suggest the growing role that science and higher mathematics were coming to play in telegraph engineering. While acknowledging his debt to Maver's work and to that of other telegraph engineers, including Fowle, Rhoads found that the available literature provided only approximations of the actual conditions and that the performance of the quadruplex and duplex was largely "a matter of guess-work, if attempted by graphics and charts."[37] To determine more accurately what occurred during transmission, Rhoads used new analytical techniques. To calculate the best arrangement of quadruplex circuits, he relied on information from oscillograph records of line current made under varying electrical conditions of wire and instruments. Interestingly, Rhoads did not present his calculations in the paper, only their results.[38]

In his earlier studies, Maver largely relied on theoretical circuit diagrams and charts in his determinations of the characteristics of quadruplex transmission. Like his contemporaries, Maver depended on experience, finding that he "was able to call up in [his] mind and to analyze the action taking place in the circuits under almost any given condition, even without graphics."[39] Indeed, Maver noted that while he did not question Rhoads's use of the oscillograph, "ordinary observation" had led earlier telegraph engineers to many of the same conclusions.[40] The graphical methods employed by Maver represented an increasing refinement of engineering design, but he and other telegraph electricians were much slower to employ the mathematical theories and analytical techniques of the emerging electrical engineering science.[41] It was in electric power engineering, and to a lesser extent in telephony, that engineering science emerged as a distinct discipline.

The changes then occurring in electrical engineering are reflected in Thomas Edison's career after he left telegraphy. His knowledge of electrical science, derived largely from the work of Michael Faraday, Joseph Henry, Georg Ohm, and other early electrical scientists, was sufficient in the early development of the new fields of electric light and power and telephony for Edison to make major contributions. By the late 1880s, however, the problems of long-distance transmission in these industries gave rise to new phenomenon that were not as easily solved by Edison's experimental approach. In long-distance telephony, Edison and other practically trained electrical inventors focused their attentions on the physical design of the transmitter in the early 1880s. But improving the efficiency and power of these instruments created new line-interference problems. It was engineers, who had graduated from the new electrical engineering programs developed by physicists at the leading American universities, who eventually solved these problems.

The new university-educated engineers were well grounded in a theoretical understanding of the physical processes of electromagnetism. By adapting the scientific methods of their professors they built new mathematical theories of electrical technology which enabled them to more readily solve these and other technical problems. The shop tradition of telegraph invention and engineering, while still adequate for the problems encountered in the mature telegraph technology, was not sufficient to solve fresh problems emerging in new electrical fields.[42] Long-time telegraph inventors and engineers, whose education came largely from a tradition of practice, were among the early leaders in electrical engineering professionalism, but their role was limited and their influence short-lived.

The pages of the industry's technical journals reflected the growing importance of telegraph engineering. In 1869, the editor of the *Telegrapher*, James Ashley, published articles complaining that "there had been little or no demand for professional electricians in this country."[43] However, within ten years, as editor of the *Journal of the Telegraph,* he could claim that

> Electrical engineering has attained the dignity of a profession, and its value and importance are being constantly more and more recognized and appreciated. Although engineers are not necessarily inventors, yet it is upon their investigations that electrical and telegraphic inventions of value must be based.[44]

Ashley's changing perspective reflected the rise of electrical departments in the telegraph companies. It also reflected attempts made by the engineers themselves to develop their profession.

The first attempts to form professional electrical engineering societies occurred in the 1870s. Early in the decade, periodic reports appeared in the telegraph press of lectures presented by prominent telegraph inventors and engineers to gatherings of operators. These lectures were an attempt on the part of professionally conscious technical personnel to provide technical training and elevate the status of operators. By the end of the decade, they were focusing on establishing formal means of communicating new scientific and technical information to other technically sophisticated inventors and engineers.

In establishing these technical societies, American telegraphers were following the example of the eminent British Society of Telegraph Engineers, to which many of them belonged as foreign members. The most prominent of these early electrical societies was the American Electrical Society, founded in Chicago in 1875. The society included many prominent inventors and engineers and its proceedings included many of the highest quality papers on electrical technology published in the United States during the five years of its existence. For reasons not altogether clear, the American Electrical Society and lesser societies founded in such cities as Cleveland, Toledo, and Cincinnati about 1878 did not survive into the new decade. The former attempted to provide a national focus for electrical knowledge, but its location away from the center of the telegraph and electrical industries in New York City may have contributed to its failure. Yet even the New York Electrical Society, which was founded in 1881 and counted most of the major electrical inventors and engineers among its members, failed to emerge as the central institution of the profession. The society eventually merged with the recently founded American Institute of Electrical Engineers. Although many of the telegraph inventors and engineers involved in the earlier societies became charter members of the Institute in 1884, the new organization achieved its success through the influence of engineers in the newer electrical industries of telephony and electric light and power. The concerns of electric-power engineers in particular dominated the business of the Institute.[45]

The growing influence of these new industries and their technical personnel was evident in the pages of Franklin Pope's journal, the *Electrician and Electrical Engineer*. Pope took over the *Electrician* in 1884 and added the new term "Electrical Engineer" to the title. The original title was a long-standing term used in telegraphy to cover a variety of telegraph employees responsible for technical operations and also to designate major inventors. The new term represented the professional consciousness emerging among electrical engineers engaged in the new electrical enterprises. For them, the label pointed forward toward theoretical and mathematical influences rather than backward toward the empirically minded telegraph electrician.[46] By 1888, when Pope dropped the original term, "Electrician," both his journal and the Institute published few articles and papers on telegraphy. Leadership within the profession had clearly passed to engineers in other electrical industries. The failure of telegraphers to play a more prominent role in the Institute indicated the extent to which the revolution in electrical engineering had passed them by.

Changes in concepts of science as well as in the nature of technological knowledge prevented telegraph engineers and inventors from remaining technical leaders in the electrical industry. The changing definition of science is illustrated by the career of Thomas Edison. When he described his Menlo Park laboratory as "filled with every kind of apparatus for scientific research" and provided the initial funding for the journal *Science,* the definition of science he was embracing was much more empirical and pragmatic than the one used in the twentieth century.[47] *Science,* which later became the preeminent American journal of science under the sponsorship of the American Association for the Advancement of Science, included articles on technologies within the electrical industry, which were helping to drive much of the research undertaken by the American physics community. To Edison and to some of the scientists with whom he associated, scientific knowledge could be gained from practical experience and experimentation. The highest value of such knowledge was in its application to the development of new inventions promoting material progress.[48]

As new technical problems not easily solved by traditional practical methods revealed the limitations of the approach relied on by Edison and his contemporaries, they also fostered the growth of university engineering programs and created a new appreciation for a more theoretical conception of science. The electrical engineers who developed and improved long-distance telephone and electric-lighting technologies did not emerge from the manufacturing shop, the telegraph operating room, the telephone switchboard office, or the central power station. Instead, they came from the nation's universities. University engineering programs were being developed by the very physicists who decried the tendency of Americans to ignore the work of scientists in their embrace of inventors. As this new generation of engineers used scientific knowledge and techniques to solve difficult technical problems, they also helped to demonstrate the value of their scientific education. Engineers in other fields were also recognizing the value of such an education and entering the classroom of American scientists. As these university-educated engineers took up careers in the corporate world, they proselytized the value of science to a wider community and used it to assert their hegemony over technical knowledge.

By the turn of the century, engineering theory rather than practical knowledge was considered the best guide to the design of new technologies. The changing character of technical knowledge was evident in the careers of two of the more famous inventors of the new century. Lee De Forest invented the radio audion tube, and Charles Kettering made significant inventions in a number of fields, most notably those developed for General Motors. Intent on a career as an inventor, De Forest attended the Sheffield Scientific School at Yale. Kettering graduated from the Ohio State University department of engineering. Their practical inclinations and interest in experimentation were not unlike those exhibited by Edison, yet both men possessed a more theoretical and mathematical bent and considered a university education necessary to develop the faculties essential to pursue invention. This was a far cry from the traditional course of practical experience and self-education that had characterized inventive careers in the nineteenth century. Though De Forest possessed a doctorate from Yale, he exhibited the same drive for independence and entrepreneurial characteristics typical of the nineteenth-century inventor; the type had not disappeared but their backgrounds were shaped by new institutions. Kettering, too, exhibited entrepreneurial characteristics, but was more typical of the engineers and Ph.D.-bearing physicists and chemists who entered industry in the twentieth century, and worked within the bureaucratic structure of the emerging corporate industrial research establishment.[49]

Incorporating Invention

The changing nature of technical knowledge and education, which helped promote a growing appreciation for the usefulness of basic scientific research, also contributed to the explosive growth of industrial research institutions in the first two decades of the twentieth century. Changing competitive conditions in American industry further fueled this growth. Antitrust legislation threatened to restrain large corporations such as Western Union in driving rivals from their markets. Western Union reacted by tolerating limited competition which did not threaten the company's dominance. Firms in other industries found that control of technology could serve as a barrier not affected by antitrust law. They also recognized that rivals controlling key technologies could also gain advantage over them. They responded defensively, by adopting research programs that took advantage of those changes in technical knowledge which placed greater reliance on scientific information and techniques in the development of new technology.

These changing conditions initially had a limited impact on telegraphy. Industry leaders were steeped in the mechanical-shop tradition of mid-century technology. But the rise of competition from the Bell system's long-distance telephone lines forced a reevaluation of telegraph technology. The technical challenges created by this competition also challenged the resources of shop-trained engineers and inventors. The decline of traditional shop invention became evident when the American Telephone and Telegraph Company (AT&T) succeeded in gaining control of Western Union. It was Western Electric engineers working for AT&T who invented the new high-speed printing telegraph systems that were adopted by Western Union. The telephone company also influenced Western Union to develop a new industrial research program.

Western Electric, which was acquired by the American Bell Telephone Company from Western Union in 1882, had begun its life as a telegraph manufacturing firm. From its beginnings it was associated with the development of new technology. One of the founding partners was telegraph inventor Elisha Gray. The company also employed skilled experimental machinists and sought the work of other inventors. American Bell purchased Western Electric in part because of this tradition of research.

By the end of the nineteenth century, the company's engineers were largely engaged in developing made-to-order switching systems and standardizing equipment rather than developing new inventions. American Bell relied instead on its own engineering division for new technology, such as the loading coil to improve long-distance transmission. The development of the loading coil in 1899 had important implications for Western Electric. It required company engineers to acquire a deeper understanding of the physics of the new device in order to improve its manufacture, and the company set up a physics-oriented research branch within its engineering department in 1911. The research branch made a major contribution to the telephone art when it developed the first electronic repeater.[50] Its example also influenced the work of other departments of the engineering department, which were responsible for developing the multiplex printing telegraph used by Western Union.

As early as 1887, one year after completing its first long-distance line between New York and Philadelphia, the Bell Telephone Company leased a private-line telegraph for use by brokers. The growth of privately leased telegraph lines in the Bell system contributed to the company's decision in 1910 to acquire Western Union. The year before, the company had begun a research program at Western Electric to develop its own page-printing telegraph, and Western Union's experience with printing telegraphy was an important consideration. Between 1910 and 1912, Western Electric engineers worked with Western Union engineers to develop what became known as the multiplex printing telegraph, designed to serve Western Union needs in handling messages on its main lines. This system used a punched tape transmitter with a five-unit code based on that of the French Baudot printing telegraph. It printed on a page form, could be quadruplexed, and allowed correction of the taped message, an important consideration for the telegraph company.[51]

The development of the multiplex printing telegraph also marked an important period of transition in the development of new inventions at Western Union. During the nineteenth century, most major telegraph inventions were conceived by individuals working as independent inventors or as contract inventors. A few were the product of company employees. For only a brief period, under the direction of William Orton, the company sought to control the direction of such inventions and to regularize research and development. In particular, Orton's support for Edison's invention factory at Menlo Park marked a faltering step toward an early form of organized industrial research.

But Orton was an unusual executive officer for a large American corporation in the nineteenth century. His close relationships to a number of inventors, and his willingness to understand not only the technical needs of the company but the technology itself gave him a much greater appreciation for the inventive process than most managers.[52] Following Orton's death, telegraph invention at Western Union largely depended upon the "enthusiastic but undirected efforts of a few employees who were responsible from time to time for the development of new ideas."[53] Although Orton did not entirely

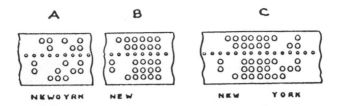

Figures 6.13, 6.14, and 6.15. The synchronous multiplex printing telegraph system developed jointly by Western Electric and Western Union engineers largely replaced the Morse telegraph system. The operating station contained a keyboard puncher on the right, a transmitter in the middle, and a page-printer at the left. The punched tape used in the system was based on a five-unit code originally developed by Frenchman Emile Baudot in 1875. The system used a time-sharing distributor similar to Patrick Delany's multiplex. (*Source:* Bell, "Printing Telegraph Systems," 201, 210.)

escape from nineteenth-century beliefs in the unpredictability of inventive genius, he nonetheless recognized that Edison himself was attempting to make invention an industrial process at his Menlo Park laboratory.

At Menlo Park, Edison used money from his contract with Western Union to support his mechanical department, the precision machine shop he had adapted from his telegraph manufactories. By adapting the machine shop solely to inventive work, Edison and his assistants could rapidly construct, test, and alter experimental devices, thus increasing the rate at which inventions were developed. In this way the laboratory became a true invention factory. The best description of the machine shop at this time is found in Edison's letter to Orton asking for financial support for it.

> I have in the Laboratory a machine shop run by a 5 horse power engine, the machinery is of the finest description. I employ three workmen, two of whom have been in my employ for five years and have

much experience. I have also two assistants who have been with me 5 and 7 years respectively both of which are very expert.[54]

A later photograph shows the shop filled with precision metalworking machines similar to those he employed as a telegraph manufacturer. Much of the machinery, as well the men to run them, came from Edison's Newark shops. Edison also drew on his telegraph background in defining problems and seeking their solutions during his inventive work at Menlo Park.[55]

In addition to serving as the principal influence on the design and operation of Edison's Menlo Park laboratory, the telegraph-shop tradition also influenced other attempts to develop forms of organized invention. Edison was not the only telegraph inventor to build a private laboratory. Patrick Delany, for example, built a laboratory in South Orange, New Jersey, not far

Figure 6.16. This illustration of Western Electric's mechanical department from an 1884 issue of *Scientific American* shows the machine shop in the background. (*Source: Scientific American* 51 [1884]: 175.)

from Edison's more famous West Orange laboratory, while Stephen Field and Franklin Pope had private laboratories at their homes. Some telegraph manufacturing shops included small laboratories for experimental work on their premises, but Edison's laboratory was a very visible model for other inventors.[56]

The influence of the telegraph-shop tradition, and of Edison's laboratory, was evident in the early telephone industry. Many of the important early improvements on telephone technology were made in the manufacturing shops, most of which initially manufactured telegraph equipment. After Bell bought Western Electric from Western Union, that shop became the principal source of both telephones and telephone improvements. To coordinate its acquisition and adoption of new inventions, the company also established an electrical and patent department in 1880, placing it under the direction of Thomas Watson, the former machinist who had become Bell's principal experimental assistant. When American Bell established an experimental shop in 1883, company officials placed it under the direction of Ezra Gilliland. A former telegraph operator, manufacturer, and inventor, Gilliland was also a long-time Edison associate and helped convince the company to hire Edison as a consulting inventor with the handsome retainer of $6,000 per year. The influence of this telegraph-shop tradition waned by the end of the century, as the Bell company turned to scientifically trained engineers in order to develop long-distance telephone service.[57] American Bell was unusual in its establishment of formal research departments to develop new inventions.

Figure 6.17. Although Bell Laboratories is more famous, especially for its contributions to physics, Western Electric's Engineering Department has played an important role in technical developments at AT&T. This photograph is of Western Electric's telephone instrument laboratory as it appeared in 1917. (*Source:* Fleming, *Industrial Research in the United States*, 30.)

In telegraphy and other industries, neither operating engineers nor manufacturing shops could be relied on to produce inventions. One of the leading telegraph engineers expressed this concern when he argued that

> in the majority of cases the active engineers in an art are fully occupied with the successful operation of existing apparatus or with problems connected therewith, and are given neither the time nor the financial aid necessary to undertake original research for the advancement of the art with which they may be associated.[58]

The development of specialized research departments such as that at Western Electric was an implicit recognition that the manufacturing machine shop could no longer serve as the central institution of invention in the electrical industry. Similar developments occurred at other major electrical manufacturers such as General Electric and Westinghouse.

The decline of research in the manufacturing shop was in part a product of the rationalization of production symbolized by the emergence of scientific management throughout American industry. Growing markets that demanded mass production and economies of scale drove this process of rationalization among the large corporations that increasingly dominated electrical manufacturing. While skilled craftsmen continued to make up a significant portion of the electrical manufacturing workforce, companies sought to eliminate much of their independence and initiative, furthering the decline of shop culture as a source of invention.[59] The expert experimental machinist of the small electrical manufacturing shop of the nineteenth century was more likely to find employment with new research institutions.[60]

Furthermore, the scale of telegraph manufacturing relative to other electrical industries was very small. The spread of telephones and electric light and power dwarfed the limited market for telegraph equipment by the end of the century, and caused many electrical manufacturers to commit the bulk of their resources to the new and growing electrical technologies. Manufacture of standard telegraph instruments other than those produced by the shops of the two operating companies accounted for less than half a million dollars by 1900, while specialized urban telegraph services such as fire alarms and district call boxes reflected a modest mass market with a manufacturing value of $1.2 million.[61] One of the few remaining manufacturers to continue producing general telegraph equipment and to remain a source of telegraph innovation was J. H. Bunnell and Company in New York, although Bunnell's death in 1899 marked a turning point for this firm as well.[62] More commonly, telegraph equipment was either a sideline for manufacturers of other electrical equipment or produced by in-house shops such as those of Gamewell Fire-Alarm Telegraph Company or the Page Machine Company, which provided printing telegraphs for Dow Jones & Company's business news service. Such in-house shops often served as research branches for these firms in much the same way that Western Electric did for the Bell system.[63]

At Western Union the company's repair shop became the basis for an electrician's workshop in 1901. Although "to all intents and purposes it was a laboratory . . . [with] the best machines and apparatus then available for experimental working in telegraph equipment," the shop was more a testing laboratory than a research facility.[64] Firms in other industries had also begun establishing similar facilities as they slowly responded to new technical challenges. Many of these laboratories, like the one at Western Union, were intended to test materials and standardize existing apparatus rather than to develop new technology. As industrial research became an important component of American industry, some laboratories did take on more significant research activities. By 1910, the inadequacy of Western Union's research support led telegraph engineers William Maver and Donald McNicol to ask "whether men thoroughly familiar with the requirements of telegraphy and of all the conditions to be met with in the technical service, and largely endowed with inventive faculty might not be selected to prosecute investigations relative to the improvement of the telegraph service as relates to the transmission of messages."[65]

When Western Union became associated with AT&T that same year, the company moved toward developing a more sophisticated research capability. As late as 1909, the engineering force of the Western Union system consisted of only ten or twelve men, "including inspectors in the field." Between 1910 and 1914, while it was a subsidiary of AT&T, Western Union's engineering department was reorganized and the telegraph company began hiring college-educated men. It also established training courses for colleges and, in coordination with Western Electric's research branch, Western Union engineers developed the multiplex printing telegraph. Two years later, after again becoming independent, the company established a research laboratory, which it housed at AT&T headquarters (formerly Western Union's headquarters). Twenty-five men worked in this new laboratory, which was enlarged in 1918 with a research and chemical laboratory, and again in 1922 with the addition of a mechanical laboratory. These men were responsible for a large proportion of the company's patents as well as for the development of important new telegraph systems.[66]

By the 1920s the scientifically trained industrial researcher was also replacing the individual inventor as a symbol of technical creativity in American society. Americans still responded positively to new technology as an agent of social progress, but they recognized that the industrial corporation had wrought profound changes in their society. Though democratic individualism, represented by the inventor-entrepreneur, had produced an expansive economy and an abundance of material goods, it also produced industrial strife and an economic order represented by the dark side of liberal individualism, the "robber barons." By 1910, political and economic forces had produced a group of industrial giants capable of exerting the kind of economic and political power which only the telegraph and railroads had possessed in the nineteenth century. Yet some prominent political reformers proclaimed the large, integrated corporation to be a "natural" product of economic evolution, as they attempted to manage it "in the best interests of society." Driven by the same belief in scientific rationalization that spurred a rising generation of new corporate managers, these reformers believed that scientific techniques could be used to manage economic and social conflict in the larger society. Because they regarded the corporation itself as a source of the technological and material progress that remained intimately bound up with mainstream American ideas of social progress, these reformers were willing to transfer their traditional faith in democratic individualism and its liberal economic and political system to the corporate economy and bureaucracy of the twentieth century.[67] In the process, the industrial researcher became an appropriate symbol for an age in which many placed their faith in the specialized knowledge of experts and which saw the large, rationally managed corporation as a model for society.

Afterword

THE FIRST decades of the twentieth century seemed to mark a sharp break with an earlier pattern of industrial research examined in this book. Instead of a range of technical personnel improving technology by using knowledge gained through practical experience and reading in technical and scientific literature, invention was now the province of formally educated scientists and engineers. Instead of taking place in informal workshops connected to manufacturing machine shops, invention now required formal laboratories. Finally, instead of focusing on improving technological products and processes, the most publicized research in the new laboratories was directed to expanding scientific knowledge that might give rise to new technology. In the public mind invention became synonymous with industrial science.

Events during the two world wars reinforced this connection. During World War I, the National Research Council mobilized the nation's academic and industrial scientists and engineers, who played a leading role in developing antisubmarine defenses, range-finders, and other military innovations that proved important after American entry into World War I. Willis Whitney, the head of General Electric's research laboratory, and John J. Carty, AT&T's chief engineer, were among the leaders of American industrial research who joined the council. After the war, the National Research Council became the principal national organization for coordinating and promoting industrial research.[1] Many of the firms establishing research laboratories in the 1920s employed scientists and engineers whose first experience with cooperative research projects on a large scale came during the war.[2] Indeed, a whole generation of industrial scientists, engineers, and corporate managers were influenced by their experience of the war. Like most Americans, they perceived the nation's entry into the war as a struggle for democracy, and, even more than their fellow citizens, they believed that industrial mobilization and technological innovation were responsible for the war's successful conclusion. The American technological superiority that had made the world "safe for democracy" thus became intimately associated with the communities of engineering and science. The leaders of American industry and its new research establishment used this association to assert claims that the future promise of American life would be found in the corporate research laboratory. Such claims were strengthened during World War II when scientific research produced important advances in radar, and most importantly, the atomic bomb. In the postwar era science has been seen as necessary for national defense just as it is for industrial strength.

The important role played by science and scientists in much recent industrial research has nonetheless obscured important continuities with nineteenth-century invention. Twentieth-century industrial research has led to its own myths, just as the independent inventors of the nineteenth century were made into heroic lone inventors. Indeed, the same circumstances that reinforced the myth of the lone inventor helped produce the new image of science-based industrial research as the prime mover in modern invention.

The heroic view of invention largely dominated histories and biographies written in the first half of the twentieth century even as the development of organized industrial research wrought a profound alteration in the nature of invention. Americans reacted to the new corporate style of invention with the same ambivalence that they expressed toward the growing dominance of large-scale organizations throughout their society. Though acclaiming the superiority of the new century's organized style of invention, they also looked back to the nineteenth century as a golden age of invention and celebrated the genius of the " 'heroic' inventor who worked picturesquely and alone in a garret."[3] Not all subsequent biographies and histories have followed this tradition, but it has dominated common perceptions of nineteenth-century invention, even though historians of technology have actively challenged such views.[4]

While historians have challenged the myth of the lone inventor, their work on corporate research has largely focused on those aspects that depend heavily on science. This has been a necessary strategy for understanding an important change in the nature of much technological knowledge and the creation of new technology. Nevertheless, by focusing on firms that were leaders in the development of laboratories in which science-based research was prominent, particularly in the electrical and chemical companies, historical research has had the unintended consequence of sustaining a belief that modern technology is little more than applied science.[5] By adopting this strategy historians have also tended to narrowly define industrial research as laboratories based primarily on scientific, academic models. Thus, one leading scholar has defined industrial research as "industrial laboratories set apart from production facilities, staffed by people trained in science and advanced engineering who work toward deeper understandings of corporate-related science and technology, and who are organized and administered to keep them somewhat insulated from immediate demands yet responsive to long-term company needs."[6]

By using this definition in his study of the early development of central research laboratories at General Electric and AT&T, Leonard Reich consciously set out to exclude categories that contemporary studies termed *indus-*

trial research. Such a definition usefully served Reich by highlighting the distinct characteristics of the form of industrial research in which he was interested, and which has been most commonly associated with the term since World War II. However, this definition neither represents the range of institutions meant by the term in the interwar period, nor captures important aspects of modern technological development that retain characteristics akin to those of nineteenth-century mechanical invention.

In its broader meaning during the first decades of this century, industrial research encompassed a variety of technological activities that gave rise to invention. While these included testing and standardization, most corporate research institutions focused directly on improving technology through minor and major improvements in products and processes.[7] Such improvements often relied little on the high-profile scientific activity performed by the more renowned corporate laboratories. Instead, a significant component of industrial research resulting in new inventions took place in engineering laboratories. Furthermore, even at those firms that created science-based central research laboratories, the development work necessary to bring major inventions to the market often entailed minor inventions and design improvements that were the product of engineering or product laboratories. For example, Bell Laboratories grew out of basic research programs developed by branches of the Western Electric Engineering Department, and this manufacturing arm of AT&T continued to perform important research functions even after such research was moved to the Laboratories. While not producing the kind of radical innovations for which Bell Laboratories became famous, the Western Electric Engineering Department continued to be an important contributor of new technology to AT&T.[8]

Most firms refined their strategies toward invention and industrial research by a process of trial and error. What set the more sophisticated strategies of companies such as General Electric, AT&T, Eastman Kodak, and DuPont apart from their contemporaries was their perception of the nature of technological change. The technical problems they faced amid challenges from competing firms convinced the managers of these corporations that deeper scientific understanding of materials and processes could lead to radical technological change. They decided to invest in long-term research programs to secure or enhance the competitive positions of their companies. Other firms took a more traditional approach by focusing their research programs on existing technology. In establishing its own form of in-house industrial research Western Union followed the model of Western Electric by placing its research laboratory under the direction of the engineering department and eschewing a program of general research in related scientific fields.

Both programs of industrial research were greatly influenced by academic sources of manpower. Even those firms that did not perceive basic science as a major source of innovation looked to scientifically trained engineers to develop new technology. This shift in the source of technical knowledge has rightly been linked to changes in the relationship between science and technology.[9] Yet, the example of telegraphy suggests another important, if secondary, factor in this transition—changing workplace relations that diminished the ability of workers to gain necessary skill and knowledge on the job. Monte Calvert was one of the first to link the decline of shop culture to new ideological as well as educational patterns in the growth of professional engineering.[10] At the heart of these changes was the managerial revolution in American corporate capitalism early in the twentieth century, which promoted an engineering rationalization of the workplace that technical and managerial personnel believed was based on a scientific foundation. The resulting transfer of technical knowledge away from the shop floor hastened the shift toward formal engineering education already underway.

While the scientific basis of the new management techniques proved largely unfounded, engineering science, the foundation of much twentieth-century technology, had deep roots in science. Edwin Layton in a seminal article described the evolution of engineering science in the nineteenth century as a "mirror-image twin" of the natural sciences. It borrowed theoretical, mathematical, and experimental methods from science but focused on very different problems. Rather than creating fundamental knowledge of the natural world, engineering theories were intended to describe the behavior of particular technologies in order to aid design practice.[11] Yet even these distinctions are blurred in the world of industrial research where new theoretical work in science has given birth to new technology and scientists often engage in what might be defined as engineering work.[12]

Although these distinctions between science and technology have become increasingly blurred, a continuing debate in engineering education suggests another important distinction between them. While engineering colleges continue to focus on the scientific and mathematical aspects of engineering knowledge, many of them have retained or reinstituted "hands-on" training programs designed to provide practical experience akin to shop knowledge. Many employers also established in-house training programs, sometimes in association with engineering colleges, designed to provide the kind of practical know-how necessary to understand the practice of technology in their industry.[13] Technology is thus different from the mirror image of science; it involves a set of skills including visual and tactile knowledge similar to that found in the nineteenth-century shop culture that is necessary for translating theory into practice.

By focusing on the rise of science-producing industrial research, historians have been able to provide only limited insights into how ideas are turned into practical technology or of the relative place of such research in American industry. While science and scientific knowledge is certainly of significance for specific industries, even in these its role may have become overstated and its character misunderstood.[14] Science continues to play its most important role in providing new information that has the potential to be used for technical purposes as breakthroughs in electrical science made possible the electric telegraph. But as occurred in telegraphy, the design of new technology and its continuing improvement can take place without new inputs from science. Although knowledge of electrical science was part of the education of inventors, practical experience with machinery and with the physical properties of electricity proved crucial to design.

Evidence from the computer industry provides a contemporary example of an industry that required basic scientific knowledge for its creation, but in which the design of new machines relies to a high degree on practical experience with the technology itself. In *The Soul of a New Machine,* Tracy Kidder provides biographical sketches of the engineers involved in the design of a new computer for Data General which indicate a common experience in their youth of taking things apart to understand how they worked—an important form of learning by doing.[15] Practical knowledge was also essential in understanding the state of the art. Kidder describes how Tom West, who conceived the idea of the new computer, obtained access to the new computer by DEC and discovered that it was not as formidable as feared by examining the machine itself. He did this by mentally taking apart the components, determining both what had been accomplished and how, and calculating its probable cost. Kidder's description suggests that modern technology retains characteristics akin to the mechanical know-how of the nineteenth century, particularly the mental manipulation of parts, be they gears and levers or integrated circuits.[16] And the design of the new machine at Data General again required practical experience. For example, one particularly troublesome bug was solved when one of the company's vice-presidents, himself an engineer, pointed out that the problem lay in the chip sockets by going over and shaking the board so that the computer failed.[17]

Kidder's book also describes a variation on shop culture in the engineering laboratory where the new machine was designed and built. The sharing of technical knowledge and experience as well as the work of building and testing new devices that lay at the heart of shop invention continue in an altered form in the shared experience of the engineering laboratory. The computer industry also contains an important example of the potential for a shop culture to emerge in a particular geographic locus. The personal computer was largely conceived and built by computer "hackers" living and working in Silicon Valley in Santa Clara County, California. Inspired by their experience with the hobbyist computer Altair, and informed through the sharing of information that took place in the Homebrew Computer Club, computer hackers in Silicon Valley designed the first "personal" computers, led by Steve Wozniak's Apple. As they turned to manufacturing, hackers found they could no longer share information in the same way and instead were forced to reform their culture within the corporate laboratories they formed at the new companies they established to exploit the new technology.[18] Just as telegraph-shop culture was shaped by corporate strategies, so too was hacker culture refashioned by the needs of the new computer corporations.

While corporate managers view the maintaining of technological secrets as a crucial component of competition, the needs of technological communities may forge lines of communication similar to the shop culture of the telegraph industry. In his recent study of innovation Erich von Hippel describes how engineers in competing firms found it useful to share what would otherwise be considered proprietary information. Von Hippel notes that "accumulated practical skill or expertise," which he terms "know-how," is often necessary for continued innovation and must be developed in-house because it is not commonly found in publications. Because such know-how, if it does not already exist at a company, is expensive to develop he found that there was an enormous incentive for engineers in competing firms who were most likely to possess such knowledge to exchange information in at least one case that he studied.[19]

Von Hippel also identified another characteristic element of nineteenth-century shop culture in the development of modern technology. While many of the innovations he discusses were the product of research and development laboratories, others relied on shop-floor knowledge gained through manufacturing and use. For example, one important innovation in pultrusion process machinery involved no formal research and development budget. Instead, the employee who made the invention "simply put equipment together out of material available on the shop floor."[20]

The link between use, manufacturing, and invention was an important one in nineteenth-century invention. Ralph Gomory, a former senior vice-president for science and technology at IBM and currently president of the Alfred P. Sloan Foundation, argues that these continue to be important in the twentieth century. He illustrates how the success of Japanese companies in the last two decades stems from their attention to these matters rather than to basic research of the type in which many American companies excel. Gomory urges American industry to follow the Japanese in paying greater attention to incremental improvements as part of a product cycle. Such improvements are intended "to refine the product, customize it for more and more consumer segments, make it more reliable, or get it to market more cheaply."[21] It was just this type of inventive activity that produced the new urban telegraph systems of the post–Civil War era and which gave rise to a period of intense technical creativity.

Another recent study, by David Mowrey and Nathan Rosenberg, also discusses the close ties between manufacturing and research and development (R&D) in many Japanese firms. They note that the engineering department of the manufacturing division of large Japanese firms continues to play a more significant role in the R&D process than do engineering departments in U.S. firms. Located at the manufacturing site, engineers in these departments acquire "considerable know-how concerning the manufacturing process," which enables them to produce the kind of rapid improvements in product design highlighted by Gomory.[22] While this system of engineering research is unlikely to produce radical inventions created by the American style of industrial research, it is most effective in producing improvements that are crucial to success in the marketplace. Mowrey and Rosenberg conclude that both Japan and the United States were moving to incorporate features of each other's research styles into their existing research and development structure in order to more effectively compete in world markets.

As modern policymakers seek to achieve a deeper understanding of the nature of innovation and the best means to encourage appropriate types of inventive activity, they may well find historical case studies a valuable source of information. Yet, while much has been written about the development and use of science-based industrial research and the radical innovations that it can produce, we still know surprisingly little about how such inventions are translated into marketable products or about the incremental inventions that continue once a product reaches the market. This history of the nineteenth-century telegraph industry may seem at first an unusual place for policymakers to turn for specific knowledge applicable to modern technology. Nonetheless, many of the patterns found in an earlier era appear to have modern counterparts. As new research is undertaken on twentieth-century technology-based R&D, similarities and contrasts with nineteenth-century patterns may well tell us much about those features in the organization of technological knowledge and activity that have application to modern invention.

Abbreviations

ADTAR   Annual Reports to Stockholders, American District Telegraph Company of Philadelphia Minutebook, WUS

ADTM   American District Telegraph Company of Philadelphia Minutebook, WUS

AT&T   American Telephone and Telegraph Company, Corporate Research Archives, Warren, N.J.

DNA   National Archives, Washington, D.C.

EC   Edison Manuscript Collection, Rare Books and Manuscripts, Butler Library, Columbia University, New York

ENHS   Edison National Historic Site, West Orange, N.J.

EP&RI   Edison Papers and Related Items, Library and Archives, The Henry Ford Museum & Greenfield Village, Dearborn, Mich.

G&SEM   Gold and Stock Telegraph Company Executive and Finance Committee Minutebooks, WUS

G&SM   Gold and Stock Telegraphy Company Stockholders and Directors Minutebooks, WUS

GP   Elisha Gray Papers, Archives, NMAH

LBG   Norvin Green, President's letterbooks, WUS

LBO   William Orton, President's letterbooks, WUS

MdSuFR   Washington National Records Center, Suitland, Md.

MM   Manhattan Quotation Telegraph Company Minutebooks, WUS

MP   Samuel F. B. Morse Papers, Library of Congress, Washington, D.C.

NjBaFAR   Federal Archives and Records Center, Bayonne, N.J.

NMAH   National Museum of American History, Smithsonian Institution, Washington, D.C.

O'RP   Henry O'Rielly Papers, New York Historical Society, New York

RL   Frank B. Rae Letterbook, Chicago office of Gold and Stock, 1887-88, WUC.

TAEB 1   *The Papers of Thomas A. Edison, Vol. 1 The Making of an Inventor (February 1847-June 1873)* (Baltimore: Johns Hopkins University Press, 1989)

TAEB 2   *The Papers of Thomas A. Edison, Vol. 2 From Workshop to Laboratory (June 1873-March 1876)* (Baltimore: Johns Hopkins University Press, 1991)

TAEM   *Thomas A. Edison Papers: A Selective Microfilm Edition* (Frederick, Md.: University Publications of America, 1985-). Citations listed by reel and frame.

VP   Alfred Vail Papers, Smithsonian Institution Archives, Washington, D.C.
WU   Corporate Archives, Western Union Telegraph Company, Upper Saddle River, N.J.
WUC   Western Union Collection, Archives, NMAH
WUDM   Western Union Board of Directors Minutebook, WUS
WUEM   Western Union Telegraph Company, Executive Committee, Minutebooks, WUS
WUS   Secretary's Office, Western Union Telegraph Company, Upper Saddle River, N.J.

Notes

Introduction

1. Telegraphy has received little attention from contemporary analysts, perhaps because it largely disappeared as a significant form of communication in the post–World War II era, just as communications studies emerged as a major field. Most studies of modern communications technology do little more than note that the development of telegraphy was a key moment in the chronology of communications history. A few also discuss it as the technical basis for the wire news services that have been prominent adjuncts of the modern newspaper, the one print media that receives significant attention in modern communications studies. A partial exception to this pattern is Beniger, in *Control Revolution,* who pays some attention to telegraphy as one in a matrix of information-processing and communication technologies developed between the 1840s and 1920s to control the technical and economic forces of the industrial revolution, and which he argues continues to influence the pattern of modern developments. Examples of modern communication studies include Dizard, *Coming Information Age;* Gordon, *Communications Revolution;* Martin, *The Wired Society;* Williams, *Communications Revolution* and *Technology and Communication Behavior.*

2. Josephson, *Edison,* 136–37. Josephson appropriately credited Norbert Wiener with this observation, but it was the biographer who brought this idea to a wide audience and who has influenced subsequent research on Edison. For statements regarding the importance of Edison's laboratory in the historical transition from lone inventors to modern industrial research see Birr, "Industrial Research Laboratories," 198; Hounshell, "Modernity of Menlo Park," 128–29; Hughes, "Edison's Method"; Hughes, *American Genesis,* 29–34; Lewis, "Industrial Research and Development," 624–25; Rae, "Application of Science to Industry," 259–60; and Reich, *Making of American Industrial Research,* 44–45.

3. Most of this literature is discussed in J. K. Smith, "Scientific Tradition," 121–31.

4. Although this pattern does not obviously apply to chemical invention, here too mechanical concerns could loom large in issues of production. See, for example, Friedel, *Pioneer Plastic,* chap. 1.

5. J. K. Smith, "Scientific Tradition." On changes in electrical engineering see especially Hughes, *Networks of Power;* McMahon, *Making of a Profession;* R. Rosenberg, "Academic Physics."

6. Edison provides the best-known example, but his closest American competitor in the race to develop an incandescent electric lighting system, William E. Sawyer, was also a telegraph inventor. Furthermore, Edison was inspired to enter the field after seeing an arc light system codeveloped by telegraph inventor Moses Farmer. And as Edison was developing his generator he investigated the arc lighting designs Charles Brush had developed in the shops of the Telegraph Supply Company in Cleveland. Telegraph shops were also the sites of the early development of telephone technology. On Edison see Friedel and Israel, *Edison's Electric Light,* and Israel, "Telegraphy and Edison's Invention Factory." No biography of Sawyer exists, but Charles Brush is discussed in Stapleton, "Early Industrial Research in Cleveland."

7. J. K. Smith, "Scientific Tradition"; Reich, *Making of American Industrial Research,* 2–3.

8. Mowery and Rosenberg, *Pursuit of Economic Growth,* 292.

Chapter 1. The Rise of American Mechanical Invention

1. H. A. Meier, "Technology and Democracy," 618.

2. For a delineation of classical republican ideology see Bailyn, *Ideological Origins;* Wood, *Creation of the American Republic;* and Shallope, "Republican Synthesis."

3. McCoy, *Elusive Republic,* 14–15, 48–69, 107–19; Wish, "Yeoman Farmer to Industrious Producer," 28–35, 64–65; H. A. Meier, "Technological Concept," 40–45.

4. Kasson, *Civilizing the Machine,* 24–27, 34–35; McCoy, *Elusive Republic,* 227–31; H. A. Meier, "Jefferson and a Democratic Technology," 17–44; Toner, "Washington as an Inventor," 313–79; York, *Mechanical Metamorphosis,* 162–63.

5. Jacob, *Cultural Meaning,* chaps. 3–7; Musson and Robinson, *Science and Technology.*

6. Jacob, *Cultural Meaning,* chaps. 5, 7; Musson and Robinson, *Science and Technology,* esp. chaps. 1–4; Middlekauf, *The Mathers,* esp. chaps. 8, 16; Greene, *American Science,* esp. chaps. 1–5; Hindle, *Pursuit of Science;* Schofield, *Lunar Society;* Robinson, "Derby Philosophical Society"; Kargon, *Science in Victorian Manchester,* chap. 1; Timmons, "Education and Technology," 135–49.

7. Hindle, *Emulation and Invention,* 14–17, and York, *Mechanical Metamorphosis,* 162–71, 183–201.

8. Hunter, *Water Power,* 41.

9. Ibid., 41–46; Hunter, *Steam Power,* 193–96; McCusker and Menard, *Economy of British America,* 322–30; Weiss and Weiss, *Early Sawmills of New Jersey* and *Forgotten Mills of Early New Jersey;* Rawson, *Little Old Mills;* Carroll, "Forest Society of New England"; St. George, "Fathers, Sons, and Identity"; Lemon, *Best Poor Man's Country,* 7–8, 146–47, 178–79, 200–207; Mitchell, *Commercialism and Frontier,* 144–47, 175; Clemens, "Chesapeake Tobacco Plantation," 527–28; de Crèvecouer, *Letters from an American Farmer,* 313–16; Hummel, *With Hammer in Hand;* Innes, *Labor in a New Land,* chap. 4; York, *Mechanical Metamorphosis,* 215–16; Fletcher, *Pennsylvania Agriculture,* chap. 13; Bidwell and Falconer, *History of Agriculture,* chaps. 9, 14.

10. McCoy, *Elusive Republic,* 76–90, 122–32.

11. Wish, "Yeoman Farmer to Industrious Producer," 72–89; McCoy, *Elusive Republic,* 223–32, 239–47; and Watts, *Republic Reborn,* 231–38, 304–11.

12. Watts, *Republic Reborn,* and Appleby, *Capitalism and a New Social Order,* particularly 79–105, provide the best analysis of this changing view, although Appleby attributes an overly capitalist outlook to the Jeffersonian Republicans of the 1790s. While some Republicans adopted a more optimistic outlook than Jefferson and pursued a liberal, capitalist vision, most were ambivalent about a society founded on self-interest. Attitudes toward manufacturing were also more ambiguous than she suggests and only as international events threatened national independence did manufacturing become more fully embraced, as was the case by 1815. Other important perspectives are offered in McCoy, *Elusive Republic;* Lemon, *Best Poor Man's Country;* Mitchell, *Commercialism and Frontier;* Banning, "Jeffersonian Ideology Revisited"; Kerber, "Revolutionary Generation"; Matson and Onuf, "Republican Empire"; Kloppenberg, "Virtues of Liberalism"; Kasson, *Civilizing the Machine,* 35–40; and Marx, *Machine in the Garden,* chap. 4.

13. Wish, "Yeoman Farmer to Industrious Producer," 92–133; Kasson, *Civilizing the Machine,* 22–31.

14. York, *Mechanical Metamorphosis,* 47–48, 185–94.

15. Article 1, Section 8, Constitution of the United States of America; Hindle, *Emulation and Invention,* 18–19; York, *Mechanical Metamorphosis,* 194–200.

16. Although the 1790 patent law was largely patterned after that of the British, Congress included a rigorous examination system in this first patent law in order to

provide greater assurance of a patent's validity than that furnished by the registration system used in Britain. Both systems allowed a limited fourteen-year monopoly to the inventor, with the right to assign all or part of the patent to others or to license its use. The American system differed most notably in providing that every patent be examined by a board of examiners, consisting of the secretary of state, the secretary of war, and the attorney general, to determine its originality and usefulness.

17. H. A. Meier, "Jefferson and a Democratic Technology," 28–32.

18. York, *Mechanical Metamorphosis,* 202–3; H. A. Meier, "Technological Concept," 306–8, 328–31, 401–2. Richard Wells, Petition on Bill to Promote Useful Arts; John Fitch, Petition on the General Patent Act; and Thomas Fiedler, Letter on Patent Laws, in Cochran et al., eds., *New American State Papers,* 4:11–19, 22. Ferguson and Baer, *Little Machines,* 6–7.

19. Preston, "Reform of the U.S. Patent Office"; Thornton, "Obtaining Patents"; S. Doc. 338 (1836); "Report of Henry L. Ellsworth to the Select Committee on the Patent Laws," *Mechanics' Magazine and Register of Inventions and Improvements* 8 (1836): 175–82; "National Academy on Patents," *Boston Mechanic* 3 (1834): 93–94; "Speculation in Patents," ibid., 163–64; "Useless Patents," ibid., 344–46; "Observations on the Patent Law," *Mechanics' Magazine and Journal of Public Internal Improvement* 1 (1830): 95–97.

20. Preston, "Reform of the U.S. Patent Office," 348–53. The problems created by the spoils system are discussed in Post, " 'Liberalizers' versus 'Scientific Men.' "

21. Willard Phillips, *Law of Patents,* 10–15, 47–52.

22. Hirsch, "Artisan to Manufacturer"; Kornblith, "Craftsman as Industrialist"; Rock, *Artisans of the New Republic,* 151–59; and Wilentz, *Chants Democratic,* 31–42.

23. Sinclair, *Philadelphia's Philosopher Mechanics,* 9–18; Wilentz, *Chants Democratic,* 40–42, 151, 272–73, 284–85; Hirsch, "Artisan to Manufacturer," 82–84. For examples of contemporary articles advocating this viewpoint see *Magazine of Useful and Entertaining Knowledge* 1 (1830): 143–47; *Boston Mechanic* 1 (1832): 72, 69–70, 72; ibid. 2 (1832): 44–46; ibid. 4 (1835): 11–12; *Mechanics' Magazine and Register of Inventions and Improvements* 1 (1833): 2–3; ibid. 2 (1833): 20, 166–67, 246; ibid. 3 (1834): 47–49; *Apprentices' Companion* 1 (1835): 1–2.

24. Hindle, *Emulation and Invention,* 127.

25. Stanley, "Patent Office Clerk."

26. In his 1981 article "Women Inventors in America," Carroll Pursell noted some of the favorable public comments on women as inventors, particularly in the pages of *Scientific American,* which was always interested in promoting its patent agency business. Nonetheless, such examples appeared infrequently.

27. See n. 24 above; Rorbaugh, *Craft Apprentice,* 33–36, 87–88. Although workingmen's journals of the 1840s, such as the *New York State Mechanic* and the *Mechanics' Advocate,* attacked wealth created at the expense of labor, they promoted new technology and advocated invention as a means of social mobility. See, for example, *New York State Mechanic* 1 (1842): 5, 21–22, 29, 45–46, 55, 75, 81–82, 97, 161, 196; and *Mechanics' Advocate* 1 (1846–47): 14, 20, 37, 46, 53, 99, 125, 132, 250–51, 298, 336, 403; ibid. 2 (1848): 140–41, 223.

28. A burgeoning literature on the development of the American working class contrasts the liberal, entrepreneurial values adopted by many masters with the classical republican values of workers. See, for example, Dawley, *Class and Community;* Faler, *Mechanics and Manufacturers;* Laurie, *Working People of Philadelphia;* Rock, *Artisans of the New Republic;* Steffens, *Mechanics of Baltimore;* and Wilentz, *Chants Democratic.*

29. Cavanaugh et al., *At Speedwell Village.*

30. Ciarlante, "Eminent American Inventors," focuses on the period 1790–1849 and

examines those inventors who have entries in the *Dictionary of American Biography*. She compares these inventors (designated DAB) to those with four or more patents but whose histories are unknown (designated NONDAB) and with the patent population as a whole. Her DAB and NONDAB inventors represent an inventive elite whose work placed them in the forefront of American industrialization.

31. On changes in traditional artisan production methods, see Wilentz, *Chants Democratic,* 107–42, and Hirsch, *Roots of the American Working Class.*

32. Laurie and Schmitz, "Manufacture and Productivity," 47–65; Laurie et al., "Immigrants and Industry," 96–105; Hirsch, *Roots of the American Working Class,* 21–36; and Wilentz, *Chants Democratic,* 111–17.

33. Quoted in N. Rosenberg, "America's Rise to Woodworking Leadership," 41.

34. On machinery in rural America see Hindle and Lubar, *Engines of Change,* 44–49, and Sloane, *Diary of an Early American Boy.* On the transition from rural to urban environments during the pre–Civil War years see Leary, "Industrial Ecology," 38–39, and Ciarlante, "Eminent American Inventors," 248–50.

35. Even as large-scale factory production in the metalworking trades became more common after 1850, skilled machinists continued to be well paid and although the chances of becoming an employer declined, skill could still be translated into upward mobility. Machinists often achieved considerable autonomy by assuming managerial functions, particularly as foremen and inside contractors, hiring their own crews and becoming responsible for the rate and quality of output. Opening a small shop also remained a possible, albeit risky, source of mobility. Gutman, "Rags to Riches Myth"; Laurie and Schmitz, "Manufacture and Productivity," 47–65; Laurie et al., "Immigrants and Industry," 103–5; Hindle and Lubar, *Engines of Change,* 181–82; Wallace, *Rockdale,* 212–13; Lozier, *Taunton and Mason,* 283–85.

36. Calvert, *Mechanical Engineer in America,* 3–6. Gibb, *Saco-Lowell Shops,* 11–12; Hindle and Lubar, *Engines of Change,* 167–84; Hunter, *Steam Power,* 172–250; Lozier, *Taunton and Mason,* 25–38; Pursell, *Early Stationary Steam Engines,* 94; N. Rosenberg, "Technological Change," 414–46; Wallace, *Rockdale,* 198–216; and White, *American Locomotive,* 13–14.

37. Hunter, *Water Power,* 172–81; Innes, *Labor in a New Land,* xviii; Lemon, *Best Poor Man's Country,* 118–19; McCusker and Menard, *Economy of British America,* 322–24.

38. Cavanaugh et al., *At Speedwell Village.*

39. See Hindle, *Emulation and Invention.*

40. Jeremy, *Transatlantic Industrial Revolution;* Pursell, "Thomas Digges and William Pearce"; Stapleton, *Transfer of Industrial Technologies;* Jeremy and Stapleton, "Transfers between Culturally Related Nations"; Hyde, "Iron and Steel Technologies," 51–58; Morison, *From Know-How to Nowhere,* chap. 2; Pursell, *Early Stationary Steam Engines,* chap. 3; N. Rosenberg, *Perspectives on Technology,* chaps. 9–11.

41. Hindle, *Emulation and Invention,* 12–14, Ferguson, "American Mechanical 'Know-How,'" 3–15; Ferguson, ed., *Early Engineering Reminiscences;* Harper, *Working Knowledge;* Scranton, "Learning Manufacture"; Danko, "Civil Engineering Practice."

42. N. Rosenberg, "Technological Change," 418–27; N. Rosenberg, *Perspectives on Technology,* 197–202; Calvert, *Mechanical Engineer in America,* 6–7.

43. Thomson, *Path to Mechanized Shoe Production;* McGaw, *Most Wonderful Machine,* 100–116, 163–70, 184; Lubar, "Corporate and Urban Contexts," 175–201; Jeremy, *Transatlantic Industrial Revolution,* 214–17, 231–38; Wallace, *Rockdale,* 198–211; Lozier, *Taunton and Mason,* 148–53; Pursell, *Early Stationary Steam Engines,* 94–113; Hunter, *Steam Power,* 172–50; Deyrup, *Arms Makers of the Connecticut Valley;* M. R. Smith, "Inno-

vation among Antebellum Arms Makers." On the role of skilled machinists also see Hoke, *Ingenious Yankees,* and Hounshell, *American System to Mass Production.*

44. Lubar, "Corporate and Urban Contexts," 175–201, and Lubar, "Nineteenth-century Pin Industry," 263–65.

45. Thomson, *Path to Mechanized Shoe Production,* describes a process by which knowledge of design was transmitted through the sale and use of machinery.

46. Hindle, *Emulation and Invention,* 15, 130, 134–35.

47. Ibid., 13–14; Preston, "Reform of the U.S. Patent Office," 340–42; "Drawings," *Boston Mechanic* 3 (1834): 31–32; "Observations on Drawing Instruments," ibid., 22–23, 33–34, 49–52, 65–68, 81–82, 97–99, 113–15, 129–32, 145–48, 169–62, 177–80; Sinclair, *Philadelphia's Philosopher Mechanics,* 38–39.

48. Munn and Company published several editions of its *Instruction on How to Obtain Letters of Patent for New Inventions* in the 1860s and 1870s. This guide, originally entitled *Hints to Inventors Concerning the Procuring of Patents,* probably appeared in the 1850s, but no extant copies have been found. Hindle, *Emulation and Invention,* 131–35; Jeremy, *Transatlantic Industrial Revolution,* 68–72; Ferguson, "Mind's Eye," 827–36.

49. See, for example, "Arts and Manufacture," *Boston Mechanic* 2 (1832): 69–70; "National Academy on Patents" (see n. 19 above); Verplanck, "Introductory Address"; and review of Dick, "On the Improvement of Society by the Diffusion of Knowledge." Henry's views are found in Mollela and Reingold, "Theorists and Ingenious Mechanics." Also see Hindle, *Emulation and Invention,* 131–33.

50. Blydenburgh, "Advancement of Useful Improvements."

51. *Mechanics' Magazine and Register of Inventions and Improvements* 7 (1836): 211.

52. See Post, " 'Liberalizers' versus 'Scientific Men' " for a discussion of this agitation for a liberal patent examination system.

53. Lubar, "Corporate and Urban Contexts," and Thomson, *Mechanized Shoe Production,* are among the few studies to analyze such communities.

54. Hindle, *Emulation and Invention,* 15–16, 18–19, 129–30.

55. Whitney, *Relationship of the Patent Laws,* 33.

56. The systematic nature of railroad technology also prompted government support but this remained largely local until the Civil War.

Chapter 2. Invention and the Development of the Telegraph Industry

1. Aitken, *Syntony and Spark.*

2. Although prominent American inventors of the first half of the nineteenth century generally had more education than the average American, it seldom matched that available to their more privileged contemporaries in Europe. Ciarlante, "Eminent American Inventors," 257–92, 333–35.

3. Bender, *New York Intellect,* 122–25.

4. Information on Morse's personal background and on his work in telegraphy is drawn from the Samuel Finley Breese Morse Papers (hereafter MP), particularly the correspondence books and scrapbooks; Morse, *Letters and Journals;* Prime, *Life of Morse;* and Mabee, *American Leonardo.*

5. Prime, *Life of Morse,* 252.

6. The relation between art and technology is the subject of Hindle, *Emulation and Invention,* and his article "From Art to Technology and Science."

7. Samuel Morse to Lucretia Walker, November 1817, quoted in Prime, *Life of Morse,* 105. For Morse's early inventions, see ibid., chaps. 4–5.

8. Cooper, "Thomas Blanchard's Woodworking Inventions," 291–97.

9. Reingold, *Papers of Joseph Henry,* 90–96; Mabee, *American Leonardo,* 190–92.

10. On the dispute see Morse, *Letters and Journals;* Prime, *Life of Morse;* Mabee, *American Leonardo;* Coulson, *Joseph Henry;* Taylor, "Henry's Contribution"; and a series of articles written by partisans of Morse, Henry, and Alfred Vail that appeared in *Electrical World* 26 (July-November 1895).

11. Agreement between Morse, Vail, Gale, and F.O.J. Smith, March 1838, Vol. 12, MP. For Gale's experiments and suggestions regarding the telegraph, see his deposition in French v. Rogers.

12. At the time of Morse's work on the electric telegraph, his brother Sidney was publishing the *New York Observer,* which he began in 1823. Previously, Sidney had published the *Boston Recorder,* a weekly religious paper.

13. Agreement between Morse, Vail, Gale, and F.O.J. Smith (see n. 11, above).

14. Vol. 12, MP. Other evidence of Vail's contribution is found in partisan accounts. These include F. L. Pope, "American Inventors," 933–35; J. C. Vail, *Early History;* Morse, *Letters and Journals,* 2:60–61, 72–73; and articles in *Electrical World* (July-November 1895).

15. Vail to Amos Kendall, 16 October 1853, quoted in F. L. Pope, "Invention of the Electromagnetic Telegraph," 181.

16. Vol. 12, MP.

17. A. Vail, *Electro Magnetic Telegraph,* 32–40, 159–66.

18. A. Vail, *Electro Magnetic Telegraph,* describes some of the changes in the instruments without indicating who was responsible for them.

19. Some evidence of the work of Vail and others can be found in letters between Morse and Vail of 19 March and 18 April 1838, 23 February 1843, 23 and 24 January 1846, and 29 August 1846, Vols. 12, 15, 21, MP. Thomas Hall describes an improvement he made to the paper-drive mechanism of the 1844 register in his testimony (p. 233), Edison Electric v. Little Electrical. Also see Henry O'Rielly to Morse, 10 March 1846, Vol. 21, MP, and excerpts from later notes written by Vail, quoted in J. C. Vail, *Early History,* 12–13, and Reid, *Telegraph in America,* 305.

20. See in particular F. L. Pope, "American Inventors," 933–34, and F. L. Pope, "Invention of the Electromagnetic Telegraph," 182–83.

21. Manuscript copy of the Report of the Sub-Committee from the Committee of Science and Arts Appointed to Examine the Electro-Magnetic Telegraph of Professor Samuel F. B. Morse, 8 February 1838, Vol. 12, MP.

22. Reid first established a telegraph system in the harbor of New York in 1821 to communicate commercial news from ships to the New York Stock Exchange. Samuel C. Reid, Memorial to Congress (26 January 1837), 24th Cong., 2d sess., S. Doc. 107; John R. Parker, Petition to Congress (1836):74, *House Journal,* 24th Cong., 2d sess.; and Reid to Secretary of the Treasury (April 1837), H. Doc. 15, 7–10.

23. H. Doc. 15 (1837), 2.

24. Agreement between Samuel Morse and Alfred Vail, 23 September 1837, Contract Book, Ser. 9, VP.

25. These systems originated in Europe during the 1790s as a means of military communication, but by the 1830s they were also used for transmitting commercial information. In the United States, the earliest semaphore telegraph was built by Jonathan Gout about 1800 between Boston and Martha's Vineyard to transmit shipping news. Other systems were later built for similar purposes in New York and Baltimore. Shaffner, *Telegraph Manual,* chaps. 2–4, and Thompson, *Wiring a Continent,* 11–12.

26. Morse to Secretary of the Treasury (27 September 1837), H. Doc. 15, 28–31.

27. H. Rept. 753 (1838).

28. John, "Origins of Commercial Telegraphy," 27–28; H. Doc. 17 (1842).

29. H. Doc. 17 (1842); John, "Origins of Commercial Telegraphy"; John, "A Failure of Vision?"

30. Of the final tally, sixty-four of the ninety "yes" votes were cast by Whigs but only five of these were from southern states, where half of the total Whig votes against the bill were cast. Southern Democrats also cast half of their party's "no" votes on the bill. Outside of the south, only New England demonstrated a wide disparity between Whig and Democrat voting, with Whigs overwhelmingly in favor and the few Democrats in the section opposed. In the mid-Atlantic states, a larger percentage of Whigs than Democrats voted favorably, but many more Democrats voted in favor or abstained than opposed the bill, while in the western states the distribution of votes between the parties was nearly equal. The controversial nature of the appropriation also led 33 percent of all Democrats and 25 percent of all Whigs to abstain, figures that were nearly the same in all sections, save the south where the percentage of abstentions by both parties was even greater. Prime, *Life of Morse*, 462–63; and List of Members of the House of Representatives, 27th Cong., 3d sess., annotated by Morse, Vol. 15, MP.

31. The register now used a steel embossing point and grooved roller to imprint on the paper in place of pencils or ink as formerly used. F. L. Pope, "American Inventors," 940–41, and A. Vail, *Electro Magnetic Telegraph*, 43–44.

32. The nature of these experiments is discussed by Morse in his 12 December 1844 report to the Secretary of the Treasury, H. Doc. 24 (1844). Prime, *Life of Morse*, 480–86.

33. Morse, *Letters and Journals*, 2:204–14; Prime, *Life of Morse*, 471–79; Mabee, *American Leonardo*, 266–68.

34. Morse, *Letters and Journals*, 2:214–22; Prime, *Life of Morse*, 479–92; Mabee, *American Leonardo*, 272–75.

35. Richard John discusses the reasons behind the economic failure of this line in "A Failure of Vision?"

36. Hurst, *Law and the Conditions of Freedom*, 6–7, 16–17, 39.

37. Goodrich, *Government Promotion*, 265–95, and Goodrich, "Internal Improvements," 145–69.

38. Prime, *Life of Morse*, chap. 10. Apart from his belief that government should specially reward his labors as an inventor, however, Morse envisioned a national telegraph system built by private enterprise but regulated by the federal government. He believed that the federal government should purchase his patents for the Post Office, which would then license private firms to build telegraph lines under government regulation. In this way a national system could be quickly built while maintaining "a system of checks" that would prevent a private company from monopolizing the invention and at the same time ensure that it was not abused by the government. Morse to F.O.J. Smith, 15 February 1838, in H. Rept. 753 (1838).

39. Hindle, *Emulation and Invention*, 16–18, 104, 130, and Phillips, *Law of Patents*, 22–25.

40. On the self-made man in America see Cawelti, *Apostles of the Self-Made Man*; Wylie, *Self-Made Man in America*; and Welter, *Mind of America*, chap. 6.

41. Welter, *Mind of America*, 149.

42. "Disappointments of the Authors of Important Improvements," *Mechanics' Magazine and Register of Inventions and Improvements* 1 (1833): 291.

43. H. A. Meier, "Technological Concept," 301–3.

44. Morse to Catherine Pattison, 27 August 1837, Vol. 12, MP.

45. On Smith's controversial relationship with Morse, see John, "Origins of Commercial Telegraphy."

46. This option was important until 1846, when government refusal to build a line to New Orleans for use in the Mexican War indicated the strong opposition to any government involvement in the telegraph and ended efforts to secure a government takeover.

47. Even though telegraph companies were incorporated in many states their charters were generally standardized (Shaffner, *Telegraph Manual*, 778–80). Hurst, *Legitimacy of the Business Corporation*, 17–30, and Seavoy, *Origins of the American Business Corporation*, 255–56, 266–68.

48. For the shift from family-owned businesses and partnerships to professionally managed corporations, see Chandler, *Visible Hand*. The emergence of the corporation as a principle vehicle of economic development and fears that it might upset the balance of economic power is described in Hurst, *Legitimacy of the Business Corporation*, 17–29, 36–43.

49. The Morse interests were divided by internal quarrels among the patent holders, while the democratic management system devised by O'Rielly failed to provide adequate coordination. A number of O'Rielly companies sought to achieve this by appointing James Reid as their common superintendent in the early 1850s, but Reid found it impossible to forge a policy upon which the individual companies could agree. Although he continually urged this alliance, known as the National Lines Company, to formally consolidate into a single company, rivalries and competing interests led to its dissolution within a few years. Thompson, *Wiring a Continent*, 39–40, 199–200, 440–41.

50. This dispute is discussed in ibid., 80–89.

51. *Dictionary of American Biography*, s.v. "House, Royal Earl"; Reid, *Telegraph in America*, 455–64; Prescott, *History, Theory, and Practice*, chap. 7.

52. Thompson, *Wiring a Continent*, 167–68, 201–2, and Reid, *Telegraph in America*, 455–65.

53. Agreement between O'Rielly and Moss, 1 December 1846, agreement between O'Rielly and Pease, 3 August 1847, and agreement between O'Rielly and Bain, 2 October 1848, all O'Rielly Papers (hereafter O'RP) 1, 1st ser.; Thompson, *Wiring a Continent*, 88, 154–55; and Reid, *Telegraph in America*, 197–201.

54. Aked, "Alexander Bain."

55. Ibid., 54–55; Reid, *Telegraph in America*, 377–80; Prescott, *History, Theory, and Practice*, 127–35.

56. S. Doc. 338 (1836); Phillips, *Inventor's Guide*; and Preston, "Reform of the U.S. Patent Office."

57. Many of these circulars, which often reprint letters to the editor and editorials from various newspapers, can be found in O'RP 1, 1st ser.

58. Mau, "Early History of the Telegraph," 248–52.

59. Ibid., 253–55.

60. Rogers also devised a modified receiving instrument. Prescott, *History, Theory, and Practice*, 159.

61. Mau, "Early History of the Telegraph," 255–56, 263–68; O'Rielly v. Morse, Decision of the U.S. Supreme Court, *Howard* 15:411–42; Thompson, *Wiring a Continent*, 194–95. Carolyn Cooper examines the role of the courts in defining invention in her dissertation, "Thomas Blanchard's Woodworking Machinery."

62. American telegraph systems operated on closed circuits in which current flowed through a line when messages were not being transmitted. This contrasted with most European telegraph systems, which used open circuits designed to prevent the flow of current when the line was not in use.

63. Prescott, *History, Theory, and Practice,* 93, 125–26, and Fischer and Preece, "Joint Report," 336–37.

64. Thompson, *Wiring a Continent,* 259–98, and Reid, *Telegraph in America,* 464–81.

65. Prescott, *History, Theory, and Practice,* 139–52, and Reid, *Telegraph in America,* 405–6.

66. Seavoy, *Origins of the American Business Corporation,* 196–99.

67. Thompson, *Wiring a Continent,* 310–30, 504–15 (the agreement).

68. Ibid., 348–70

69. Hall, *Organization of American Culture,* 240–70, and Trachtenberg, *Incorporation of America,* 3–7.

70. Chandler, *Visible Hand,* 8.

71. Ibid.; Hughes, *Networks of Power;* Hughes, "Evolution of Large Technical Systems"; Mayntz and Hughes, *Development of Large Technical Systems;* Usselman, "Running the Machine."

Chapter 3. Invention and the Telegraph Technical Community

1. See chapter 1 for a discussion of emulation. On shop culture, see Calvert, *Mechanical Engineer in America.* Calvert coined the term *shop culture* to describe the nature of the technical community formed in those industries, which served as the birthplace of mechanical engineering. In similar fashion, early electrical engineering may be said to have emerged from the telegraph industry in the form of a shop culture concentrated in the telegraph operating room and the urban telegraph manufactory.

2. One telegraph engineer noted that on a line of five hundred miles there would be nearly fifteen thousand insulators. If poorly designed, they would easily dissipate much of the current by allowing it to escape into the ground. Insulation continued to be difficult for several decades. Attempts to solve this problem accounted for the greatest number of patented telegraph inventions of the prewar period, even though some attempted improvements were never patented, and the problem continued to plague lines into the postwar period. Prescott, *History, Theory, and Practice,* 262–69; Shaffner, *Telegraph Manual,* 529–58; Reid, *Telegraph in America,* 122–23, 127, 156–57, 169–70, 190, 279–80, 355–56, 415, 459, 504–8, 648–50; Fischer and Preece, "Joint Report," 246–54.

3. Reid, *Telegraph in America,* 127–28, 310, 647–48, and Prescott, *Electricity and the Electric Telegraph,* 260–61.

4. *Dictionary of American Biography,* s.v. "Rogers, Henry J."; depositions of John S. McCrea and Charrick Westbrook in French v. Rogers; Shaffner, *Telegraph Manual,* 370–72; Samuel Bishop advertisement, *Telegrapher* 1 (1864–65): 50.

5. Reid, *Telegraph in America,* 37–75; *Dictionary of American Biography,* s.v. "Farmer, Moses Gerrish"; Hounshell, "Inventor as Hero." Unfortunately, the Moses Farmer Papers in the Department of Special Collections, University Research Library, University of California–Los Angeles, shed little light on his early work in telegraphy.

6. Thomas Hall's testimony, Edison Electric v. Little Electrical, 237; Fischer and Preece, "Joint Report," 282, 327–28, 337–39, 447–48; Shaffner, *Telegraph Manual,* 463–64, 752–53.

7. Reid, *Telegraph in America,* 122, 647, and obituary of David Brooks, *Journal of the Franklin Institute* (1891): 75–77.

8. Reid, *Telegraph in America,* 106, 119–20, and Shaffner, *Telegraph Manual,* 570.

9. Shaffner, *Telegraph Manual,* 701–3, 752; "Regulations," Magnetic Telegraph Company Minutes 1:12–14, Box 3369, WU.

10. "To the Telegraphic World," *American Telegraph Magazine* 1 (1852): 1.

11. "Practice Illustrative of Theory," ibid., 21.

12. Reid, *Telegraph in America,* 666–70.

13. Shaffner, *Telegraph Manual,* 762–63.

14. Prescott, *History, Theory, and Practice,* 1.

15. Gabler, *American Telegrapher,* 50–52, 68–71, and Abernethy, *Modern Service,* 194–98.

16. "Progress of Business Connected with the Telegraph, &c." *American Telegraph Magazine* 1 (1852): 21, 27; Chester's deposition in French v. Rogers, 349–50; "Our Advertisers," *Journal of the Telegraph* 1 (1868): 5; obituary of John Chester, ibid. 4 (1871): 264; obituary of Charles Chester, ibid. 13 (1 May 1880): 136.

17. Davis's *Manual of Magnetism* appeared in three editions between 1842 and 1851. Post, *Physics, Patents, and Politics,* 9, 18, 29, 37, 40; Newhall, *Daguerreotype in America,* 39, 120, 143; Jenkins, *Images and Enterprise,* 22; Thomas Hall's testimony, Edison Electric v. Little Electrical, 218, 233.

18. On Boston manufacturers see listings under "Instruments," *Boston City Directory;* "Death of the Oldest Telephone Manufacturer (Charles Williams, Jr.) 1908," Box 1071, AT&T; T. Hall, *Illustrated Catalogue,* 7–8, and Hall's testimony, Edison Electric v. Little Electrical, 217–53.

19. *Dictionary of American Biography,* s.v. "Clark, James Johnson."

20. Reid, *Telegraph in America,* 640–42, and F. L. Pope, "Western Union Telegraph Company's Manufactory," 104–5.

21. "Progress of Business Connected with the Telegraph, &c." (see n. 16 above), 21–22.

22. Advertisement for Norton's Telegraph Rooms, ibid, endpage.

23. Information concerning inventive work by manufacturers prior to the Civil War is difficult to find, but scattered references appear in Prescott, *History, Theory, and Practice,* 26–33, 178; Reid, *Telegraph in America,* 303–4, 656; Shaffner, *Telegraph Manual,* 102–4, 446–55; and Moreau, "Story of the Key." In his testimony in Edison Electric v. Little Electrical (p. 237), Thomas Hall claimed that "in about the first of the War or before it, there was [sic] electrical establishments in all the cities and towns of the United States. They kept materials for telegraph offices and jewelers used to keep them—[also] hardware stores."

24. Prescott, *History, Theory, and Practice* (1866), 459–77; Reid, *Telegraph in America,* 145–46, 150, 324–25, 650–51; *TAEB* 1, 30–31, 40–47, 114–15.

25. By the mid-1870s when Grandy became experienced enough to devise "two or three other forms of repeater which worked successfully and differed more or less from the conventional forms," he found that there was little demand for new repeater designs and neither took out patents nor sought to interest one of the telegraph companies in his design (letter to the editor, *Telegraph Age* 14 [1894]: 42). Grandy later patented an improvement in duplex relays (U.S. Pat. 230,001).

26. *TAEB* 1, Docs. 14–22, 32, 59.

27. Prescott, *History, Theory, and Practice,* 81–84; Shaffner, *Telegraph Manual,* 446–51; Reid, *Telegraph in America,* 650; *TAEB* 1, Docs. 11–13, 30, 40.

28. *Dictionary of American Biography,* s.v. "Stager, Anson"; Shaffner, *Telegraph Manual,* 102–4, 446–55, 492–95, 837–39.

29. My thinking on community has been stimulated by Thomas Bender's challenging essay, *Community and Social Change in America.* Some general works dealing with the emergence of translocal communities and organizations in nineteenth-century America include Wiebe, *Search for Order, 1877–1920;* Brown, *Transformation of American Life;* and P. D. Hall, *Organization of American Culture.*

30. Some of the operators' lore and anecdotes were collected together in the following works: Huntington, *Telegrapher's Souvenir;* Johnston, *Telegraphic Tales;* and Walter Phillips, *Oakum Pickings.* Others are readily found in the pages of telegraph journals such as the *Telegrapher,* the *Journal of the Telegraph,* and the *Operator.*

31. As in the railroad industry, the establishment of formal bureaucratic organizations did little to regularize labor relations. In part this appears to be have been the product of their geographically decentralized operations. For a discussion of labor relations in the railroad industry, see Licht, "Dialectics of Bureaucratization" and Licht, *Working for the Railroad.*

32. Gabler, *American Telegrapher,* 146–58, and Ulriksson, *Telegraphers,* 15–30.

33. A general overview of differences in telegraph practice and personnel among the United States, Great Britain, France, and Germany is found in Israel and Nier, "Transfer of Telegraph Technologies." On France see Butrica, "Genesis of Electrical Engineering," 365–67; Butrica, "From *Inspecteur* to *Ingénieur*"; and Shaffner, *Telegraph Manual,* 773–76. No comparable studies exist for either Britain or Germany. The creation of the Society of Telegraph Engineers in 1871 indicates the well-developed state of the telegraph engineering profession by the time Britain nationalized its telegraph system. For some idea of the training given British telegraph operators see Preece, "On the Advantages of Scientific Education." For the history of the Society of Telegraph Engineers see Reader, *Institution of Electrical Engineers,* and Appleyard, *History of the Institution of Electrical Engineers.* In Germany state educational systems provided training for government technical services. For engineers in state service, formal technical education was required. In many of the German states courses and examinations covering electrical telegraphy were established. Seidel, "Von der electrischen Telegraphie zur Elektrotechnik," and Jungnickel and McCormmach, *Intellectual Mastery of Nature,* 217–18, 255.

34. Some telegraphers did express concern over the state of telegraph engineering in America and undertook to improve engineering practice. See, for example, Prescott, *History, Theory, and Practice,* 1–2, 262; Shaffner, *Telegraph Manual,* 514–15, 519–23; and Reid, *Telegraph in America,* 426.

35. On *Scientific American,* see Borut, "*Scientific American.*" A number of telegraph operators, manufacturers, and inventors read *Scientific American* as evidenced by letters and articles they published in the journal.

36. Over the years proposals that telegraph companies establish libraries to assist the acquisition of such knowledge failed to generate much enthusiasm. As early as 1849, while acting as superintendent of Magnetic Telegraph Company lines, James Reid had recommended that the company furnish "a library of scientific works, chiefly electric, for the information of your operators, at each office" ("Proceedings of the Convention of Delegates and Board of Directors Representing the Pittsburgh, Cincinnati and Louisville Magnetic Telegraph Company, held in the City of Columbus, Ohio, June 6th and 7th, 1849," O'RP 2:8, 1st ser.). When the National Telegraphic Union was founded in 1864, a proposal to establish a library in the Western Union offices in New York received encouragement from Samuel Morse and other prominent figures but was not carried out, and ten years later a similar proposal also failed ("A Telegraphers' Library and Reading Room, *Operator* 2 [1 December 1874]: 4).

37. Fischer and Preece, "Joint Report," 445–46.

38. Some operators did receive training in telegraph schools, but most had a deservedly bad reputation. One of the more promising schools was established in Oberlin, Ohio, at the Calkins, Griffin & Co.'s Business Institute in 1866. Among the members of its advisory board were Jeptha Wade, then president of the Western Union Telegraph Company, and John Caton, president of the Illinois and Mississippi Telegraph Company. Other telegraph

officials were also on the board and the school itself was under the direction of Chester H. Pond, an experienced operator who later had several inventions to his credit. Among the instructors were two professors from Oberlin College, and the school promised to teach its students theoretical telegraphy. By 1870, however, the reputation of the Oberlin school was no better than most. Pond, *Outline of Theoretical Telegraphy;* "A Telegraph College Graduate," *Telegrapher* 6 (1869–70): 264. For later comments on the problem of telegraph schools see "Telegraphic Colleges" and "Students," both in *Operator* 12 (1881): 485–86, 490.

39. Fischer and Preece, "Joint Report," 316–28, 603.

40. Seiler's testimony, Ladd v. Seiler, 37.

41. Watson, *Exploring Life.*

42. Haskins's testimony, Edison Electric v. Little Electrical, 392.

43. Welch, *Charles Batchelor.*

44. Abernethy, *Modern Service,* 44–45; *TAEB* 1, Doc. 7; Reid, *Telegraph in America* (1886), 764.

45. "Reasons Why," *Telegrapher* 1 (1864–65): 22, and "The National Telegraph Union," ibid. 2 (1865–66): 93.

46. "Permanent Enlargement of the Journal. How to Increase Its Circulation," *Journal of the Telegraph* 6 (1873): 41, and Reid, *Telegraph in America,* 535–36.

47. "Practical Suggestion," ibid. 3 (1870): 254.

48. "Suggestions to the Telegraphic Fraternity," ibid. 11 (1878): 20.

49. *Operator* 16 (1885): 89. Other papers concerned solely with telegraphy replaced the *Operator.* These papers, such as *Telegrapher's Advocate* and *Telegraph Age,* served largely as fraternal publications. The latter, however, was later edited by Johnston, who began to introduce more technical articles. It eventually became *Telegraph and Telephone Age,* following the brief takeover of Western Union by AT&T. Even prior to that, however, it had been running articles on telephony.

50. "The Improvement of Telegraphists and the Telegraphic Service," *Journal of the Telegraph* 11 (1878): 230.

51. Abernethy, *Modern Service,* 42–43. Abernethy's book appeared in several revised versions between 1882 and 1904.

52. On Prescott see *Dictionary of American Biography,* s.v. "Prescott, George Bartlett," and Reid, *Telegraph in America,* 357–58, 384, 409, 421, 563–64, 657. On Stager see *Dictionary of American Biography,* s.v. "Stager, Anson"; Shaffner, *Telegraph Manual,* 837–39; Reid, *Telegraph in America,* 161, 187–88, 469, 480, 483, 536, 561–62; and correspondence with William Orton in LBO. On Van Horne see Reid, *Telegraph in America* (1886), 667–68, and Taltavall, *Telegraphers of To-day,* 14–19.

53. Gabler, *American Telegrapher,* 108–9.

54. Ibid., 50–51, 71–72, 116–21, 124–25, 134–35. Pay scales for women operators generally ranged between twenty-five and sixty dollars per month, with a few receiving salaries as high as seventy to ninety dollars. Most earned thirty to forty dollars. Wages for male operators were generally twice those of women.

55. Orton to J. W. Phelps, 13 February 1869, LBO 5:263–66. Telegraph schools, such as Cooper Union, and business schools offering courses in telegraphy remained important sources of education for female operators. In large city offices women, especially daughters of the Irish working class, also found employment as check-girls and followed more traditional apprenticeship routes into the profession. Gabler, *American Telegrapher,* 112–13.

56. Andrews, "What the Girls Can Do."

57. Orton to J. W. Phelps, 13 February 1869, LBO 5:263–66.

58. Andrews, "What the Girls Can Do," 111–12, and Gabler, *American Telegrapher*, 108–11.

59. Gabler, *American Telegrapher*, 112.

60. Andrews, "What the Girls Can Do," 117.

61. Gabler, *American Telegrapher*, 111. One of the first female operators worked out of the sitting room in her home, filling the time between messages "fixing her Sunday bonnet, or 'embroidering articles of raiment'" (Andrews, "What the Girls Can Do," 116).

62. "Dr. Bradley's Clock and Works," *Journal of the Telegraph* 3 (1869): 5, and Scribner, "Log-Fire Reminiscences," 4. Some understanding of the skill and patience involved in winding coils is suggested by the experience of Jane Leigh and Susan Perrett, who discovered during their attempts to reconstruct the coils Faraday used in his induction experiments that it took two people around ten hours (Gooding, "In Nature's School," 132 n. 16). In *Electrical Workers* (p. 31), Ronald W. Schatz notes that coil winding for motors and dynamos was also "intricate, painstaking work."

63. The patent infringement case involving Noyes's patents (U.S. Pats. 359,687 and 369,688) could not be located before publication. Municipal Signal v. Gamewell, *Federal Register* 52:464. No biographical information could be found to determine with certainty that Noyes was female.

64. Brinkerhoff and Cumming first became associated about 1876, when they were both listed in the New York City Directory as residing at 303 E. 19th Street. Clara was listed as a music teacher and George as a clerk. He apparently gave up telegraphy about 1876, the first year he appears in the directory and the only year he is listed as a telegraph operator. For Cumming and Brinkerhoff see their descriptive letterhead on a letter to Edison dated 3 March 1884 (*TAEM* 71:69). Edison also exhibited the telegraph key as part of his exhibit at the Paris Electrical Exhibition of 1881. Letters regarding that exhibition and a circular for "Cumming's Periphery Contact Telegraph Key" are found in *TAEM* 58:981, 1095, 1097, and 1108. Cumming assigned Brinkerhoff a half-interest in his patent in May 1882 (U.S. Pat. 256,645). The Cumming-Brinkerhoff contact-point patent was U.S. Pat. 256,646.

65. Information about these firms can be gleaned from advertisements appearing in the *Telegrapher* and the *Journal of the Telegraph* and from articles that appeared on the following pages of those journals: *Telegrapher* 4 (1867–68): 99; 5 (1868–69): 17, 302, 399, 413–14; 6 (1869–70): 145–46; 8 (1871–72): 104–5; *Journal of the Telegraph* 1 (1 June 1868): 4; 2 (1869): 5, 43, 212, and 15 March 1869 suppl.; 3 (1870): 33, 96, 233; 5 (1872): 215. Also see testimony appearing in the patent interference cases cited elsewhere, and Fischer and Preece, "Joint Report," 415–29.

66. Taltavall, *Telegraphers of To-day*, 223–24.

67. Ibid., 48.

68. For a comparison of commercial and railway telegraphy, see Abernethy, *Modern Service*, and an anonymous account published under the pseudonym Samson, *Sam Johnson*, 172–73.

69. Railway telegraphers used the basic Morse key and sounder system, but over time a separate technical system of automatic railway signaling also developed as a nearly independent branch of telegraphy. Most of those involved in railroad signaling technology did not generally participate in the development of other telegraph technology and vice versa, although there were a few individuals whose inventive work overlapped. For a discussion of the development of railroad signaling technology see Usselman, "Running the Machine," chap. 2.

70. U.S. Pat. 90,270; Orton to Jones, 29 May 1868, LBO 3:464; Reid, *Telegraph in America,* 322; "The Association of Railroad Telegraph Superintendents," *Journal of the Telegraph* 16 (1883): 49.

71. Taltavall, *Telegraphers of To-day,* 57, 204, and U.S. Pats. 230,001, 252,346, 286,107, and 324,799.

Chapter 4. The Urban Technical Community and Telegraph Design

1. Seavoy, *Origins of the American Business Corporation,* 196–99, and Carosso, *Investment Banking in America,* chaps. 1–2.

2. DuBoff, "Telegraph and the Structure of Markets," 266–69.

3. The annual number of U.S. patent applications nearly doubled from 10,664 in 1865 to 21,276 in 1867, which was nearly three times the number in 1860 (7,563). U.S. Bureau of the Census, *Historical Statistics,* 8, 959.

4. The transition from rural to urban environments during the pre–Civil War years is discussed in Ciarlante, "Eminent American Inventors." Thomas P. Hughes notes the importance of such urban manufacturing centers as well as connections to a city's financial community in discussing inventor Elmer Sperry's long stay in the city of Cleveland in *Elmer Sperry* (p. 70) and "Professional Inventor in the Heroic Age."

5. This learning process is demonstrated in Thomson, *Path to Mechanized Shoe Production.*

6. N. Rosenberg. "Technological Change"; Pred, *Spatial Dynamics,* 24–31; R. L. Meier, "Organization of Technical Innovation"; Higgs, "Urbanization and Inventiveness"; Lubar, "Corporate and Urban Contexts," chap. 4; Lubar, "Nineteenth-century Pin Industry."

7. Most telegraph operators, who made up the population from which the majority of telegraph inventors emerged, worked in small towns and villages, or at small railroad depots. Most in fact began as railroad operators. Gabler, *American Telegrapher,* 59, and Abernethy, *Modern Service,* 41.

8. Most major telegraph inventors found their way to New York City, the nation's financial capital and also the center of the telegraph industry. The dominant role of New York City is evident from those inventors with five or more patents, who therefore presumably sought to make invention an important part of their careers. Fifty-three percent of such inventors worked in the New York metropolitan area (which included parts of New Jersey, Long Island, and towns immediately to the north of the city), while only in the Boston, Philadelphia, Chicago, and Cincinnati metropolitan areas were there more than two inventors with five or more patents. Even inventors with few patents found New York City an important center and 26 percent of those with fewer than five patents came to the city. Only Boston, with just over 6 percent, Philadelphia, with just over 4 percent, and Chicago, with just over 3 percent, could claim significant numbers of inventors with fewer than five patents. However, over fifty other cities and towns had at least two inventors, most of whom were working at about the same time and who occasionally coinvented. These statistics are based on a survey of U.S. patents granted between 1866 and 1890 (see n. 36 in this chapter, below).

9. Orton to Flanery, 29 July 1873, LBO 12:178–85.

10. See, for example, Flanery's letters to the editor, *Journal of the Telegraph* 9 (1878): 150–51; 12 (1879): 5.

11. *TAEB* 1, Docs. 34, 41, 44, 65, and chap. 3 introduction.

12. "Bishop's Telegraph Rooms," *Telegrapher* 4 (1867–68): 335.

13. "Electrical and Telegraphic Manufactures in Philadelphia," *Telegrapher* 5 (1868–69): 17.

14. Watson, *Exploring Life,* 52–59; deposition of Ernest P. Warner, Edison v. Lane v. Gray v. Rose v. Gilliland, 5; "Catalogue of the Western Electric Manufacturing Company" (1876), ENHS.

15. See advertisements appearing in *Operator* 4 (1875–76). Gilliland's address was New York, but his shop was actually located within Thomas Edison's larger shop in Newark, N.J.

16. *TAEB* 1, Doc. 68.

17. Paul Seiler's testimony, Ladd v. Seiler, 38.

18. This pattern is explored in Friedel and Israel, *Edison's Electric Light.* Charles J. Wiley's testimony, Wiley v. Field, 39–41.

19. George M. Phelps's testimony, Phelps v. Anders (3993), 41.

20. "Electrical and Telegraphic Inventions," *Journal of the Telegraph* 13 (1880): 292. Similar sentiments were expressed privately by inventor Elisha Gray in a letter to A. L. Hayes of 7 May 1875, Box 2, Folder 1, GP.

21. Charles E. Buell's testimony, Buell v. Martin, 8.

22. Orazio Lugo's testimony, Field v. Lugo.

23. *TAEB* 1, Appendix 1, A19–23, D219–20; Henry C. Nicholson's testimony, Nicholson v. Edison; Jones, "Quadruplex," 26; Taltavall, *Telegraphers of To-Day,* 44–45, 265–66; "The Napoleon of Science," *New York Sun* (10 March 1878) in Charles Batchelor Scrapbook, Cat. 1240, *TAEM* 94:119–20; Emile Shape's testimony, Western Union v. B&O. Haskins, it should be noted, was general superintendent of the Northwestern Telegraph Company lines on which his duplex was tested.

24. Summaries of patent assignments can be found in the Patent Assignment Digests, U.S. Patent Office Records, DNA, and copies of the assignments themselves are found in the Patent Assignment Libers of the U.S. Patent Office, which are located at MdSuFR.

25. Charles J. Wiley's testimony, Wiley v. Field, 46.

26. Jesse H. Bunnell's testimony, Cochrane v. Van Hoevenbergh, 22.

27. G. D. Smith, *Anatomy of a Business Strategy,* 83.

28. Affidavit of Enos M. Barton, LaRue v. Western Electric.

29. *Rules of Practice in the United States Patent Office* (Washington, D.C.: Government Printing Office, 1875). Some idea of standard attorney fees can be gleaned from the bills of Edison's patent attorney, Lemuel Serrell, found in folders 75-008, 76-012, 77-015, 78-028, *TAEM* 13:335, 1019; 14:470, 623.

30. William B. Watkins's testimony, McCullough v. Watkins, 70; Edison's U.S. Patent Office application files, U.S. Patent Office Records, MdSuFR (a microfilm of this is located at the office of the Edison Papers, Rutgers University).

31. William Watkins, for example, was fortunate that his wife, Henrietta Watkins, was able to provide funds to help him meet his experimental expenses (Watkins testimony, McCullough v. Watkins, 69). Several inventors assigned patents to individuals with the same last name, suggesting that these were relatives.

32. Correspondence between Mary E. Odell and Edison, 19 and 25 August 1883, and L. Pulliam to Edison, 2 January 1884, *TAEM* 64:413, 1202, and 71:390.

33. For documentation of Edison's career in Boston see *TAEB* 1, chaps. 2–3.

34. Welch continued to support George Anders's later inventions and together they established a thriving business to manufacture and supply private-line telegraphs. Anders v. Warner; Phelps v. Anders (3993 and 6739).

35. For Edison's early career in the New York area see *TAEB* 1, chaps. 3–11.

36. Much of this inventive activity took place in New York, the nation's largest city and telegraph center; 24 percent of the inventors patenting urban systems worked in the New York metropolitan region, which also boasted 27 percent of all telegraph inventors.

No other city rivaled it. Boston was closest with 7 percent of the total and 8 percent of urban telegraph inventors; Philadelphia could claim 5 and 4 percent; Chicago had about 3 percent of each, as did Washington, D.C., and Baltimore combined. Cincinnati was the only other city with as many as 2 percent of the inventors working on urban systems.

The period included in the survey of telegraph inventors and their patents was 1866–90. There was a total of 922 inventors, 496 of whom worked in categories related to urban telegraphs. These categories included printing and dial telegraphs used for private lines and market reporting, district telegraphs, and central station systems for burglar, fire, and police alarms. The percentage of urban telegraphs would be even higher if such items as burglar and fire alarms for single buildings, hotel annunciators, and watchman signals were included. A total of 115 inventors (half of whom also worked on urban telegraphs) designed inventions for increasing message density on long-distance lines, and 368 inventors worked only on components. Inventors working only on railway telegraphs were not included; for a discussion of railroad signaling technology see Usselman, "Running the Machine," chap. 2.

37. The strong mechanical characteristics of urban telegraphs are evident not only from patents, but also from testimony and exhibits in patent interference cases cited in this chapter. Material in *TAEB* 1 also makes apparent the mechanical nature of urban telegraph technology. Also see Maver, *American Telegraphy*, chaps. 25–29.

38. The following discussion of the development of urban telegraph markets draws on Rosenberg and Israel, "Intraurban Technology"; Tarr et al., "The City and the Telegraph"; and Reid, *Telegraph in America*, 596–636.

39. Testimony of Robert Hoe on behalf of Morse, 6 and 8 May 1854, Morse patent extension case, U.S. Patent Office Records, DNA; Reid, *Telegraph in America*, 596.

40. *TAEB* 1, Doc. 41.

41. Reid, *Telegraph in America*, 621–25.

42. The best information concerning the technical development of private-line printing telegraphs is found in *TAEB* 1.

43. Ladd to Welch, 22 October 1875, Phelps v. Anders (6739).

44. Charles S. Jones to Welch and Anders, 13 April 1874, Phelps v. Anders (6739). For additional reactions to Anders's printer see the other letters entered in evidence in this case.

45. On the use and development of printing telegraphs see Reid, *Telegraph in America*, 602–15; Fischer and Preece, "Joint Report," 377–83, 502–11; Healy, "Stock Tickers"; Hotchkiss, "Stock Ticker"; Calahan, "Evolution of the Stock Ticker"; R. W. Pope, "Stock Reporting Telegraph." The continuing importance of mechanical considerations for printing telegraph technology in the early twentieth century is described in the testimony of John M. Joy in Page Machine v. Dow Jones.

46. Henry Bentley set up the first such system in New York City in the mid-1850s with local offices throughout Manhattan, Brooklyn, and Williamsburg, and later established a similar service in Philadelphia. Bentley's Philadelphia Local Telegraph Company later became the American District Telegraph Company of Philadelphia. Reid, *Telegraph in America*, 598–600.

47. On district telegraphs see Calahan, "District Telegraph"; Greer, *Alarm Security*, 21–30; and Gibson and Lindsay, "Electric Protection Services," 21–28.

48. The telephone was conceived by inventors working on multiple telegraphy and was first introduced in an urban context growing out of the early district telegraph systems. Reid, *Telegraph in America*, 634–36; Greer, *Alarm Security*, 28–30; Tosiello, *Birth and Early Years,* chap. 5.

49. The present-day alarm company ADT, for example, is the descendent of the

nineteenth-century American District Telegraph Companies. ADTAR; Gibson and Lindsay, "Electric Protection Services," 21–28; Greer, *Alarm Security*, 28–30, 50–59.

50. "Brief on Behalf of Ruddick," Ruddick v. Milliken.

51. Tarr et al., "City and the Telegraph," 13; Gibson and Lindsay, "Electric Protection Services," 11–12; Greer, *Alarm Security*, 13–22; von Fischer-Treuenfeld, "Fire Telegraphs," 75–121; von Fischer-Treuenfeld, "Present State of Fire-Telegraphy," 258–307; Examiners Report, 8 May 1871, Channing-Farmer Extension Case, Box 59, U.S. Patent Office Records, DNA; Teaford, *Unheralded Triumph*, 240–45.

52. Ebenezer S. Crocker's testimony, Phelps v. Anders (3993); Chester, *American System*; Flinn, *Chicago Police*, 397–407; Gamewell Fire-Alarm Telegraph Company, *Police Telephone*.

53. Reid, *Telegraph in America*, 644–45.

54. In his letter of 24 January 1871 to Anson Stager, William Orton described Phelps and Edison as the leading "electro-mechanicians" in the telegraph industry (LBO 8:317–21). Henry Fischer and William Preece referred to Phelps as "one of the most eminent electrical mechanicians of the day" in their "Joint Report," 558. Thomas Watson, whose primary inventive work came in telephony, claimed that his success as an inventor and electromechanician was due to his constant study of ways to eliminate useless mechanical movements (Watson, *Exploring Life,* 34–35).

55. Notebook of telegraph escapements drawn in February 1872, Box 13, Item 1, EP&RI.

56. The table of contents and preface of Edison's book are *TAEB* 2, Docs. 408 and 409. See n. 3 in the headnote preceding Doc. 408 for the location of chapter drafts and notes.

57. For his quadruplex designs see *TAEB* 1, Docs. 301 and 387, and *TAEB* 2, Docs. 348, 387, 469, and 488. For one of his mechanical acoustic designs see *TAEB* 2, Doc. 664.

58. *TAEB* 1, Doc. 101.

59. See, for example, *TAEB* 1, Docs. 180 and 194. Designs for mechanical perforators are scattered throughout *TAEB* 1; also see *TAEB* 2, Docs. 349 and 556.

60. *TAEB* 2, Docs. 458 and 615.

61. Gray to Hayes, 7 May 1875, GP.

62. *TAEB* 1, Docs. 11–22, 40, 59; copy of circular for "Pond and Gray's Patent Self-Adjusting Relay," agreement between Gray and Chester H. Pond, 10 October 1867, and correspondence between Gray and his wife, 7 August, 6 and 21 November 1867, GP; "Gray's Shunt Repeater," *Telegrapher* 7 (1870–71): 289.

Chapter 5. Invention and Corporate Strategies

1. It should be noted that interconnection between local and long-distance telegraph services were primarily achieved by administrative rather than technological means.

2. At the turn of the century, Western Union would form a single American District Telegraph Company out of the formerly autonomous local companies. Gibson and Lindsay, "Electric Protection Services," 28; Greer, *Alarm Security*, 54; Reid, *Telegraph in America* (1879), 633–34; ibid. (1886), 804–7; testimony and exhibits in Ladd v. Seiler and Wilson v. MacKenzie.

3. *TAEB* 2, Docs. 415, 447, 510–11, 545, 553, 582, 614–15, 618 n. 7, 653–54, and introductions to chaps. 3, 4, 7, 8, and 10.

4. For example, American District of Philadelphia, which switched from Western Union to Atlantic and Pacific, found its revenues from general telegraph business (i.e.,

messages transferred to long-distance companies) were seriously affected after Atlantic and Pacific was taken over by Western Union. Later, the company found that competition between Western Union, Postal Telegraph, and Baltimore and Ohio Telegraph led to competitive bidding for interconnection. ADTM, 177, 199, and ADTAR, 20 May 1878 and 16 May 1887.

5. Wilson's title with American District was vice-president, but his salary was only twenty dollars per month, suggesting that the title meant very little.

6. Testimony and exhibits, Wilson v. MacKenzie.

7. Seiler's testimony, Ladd v. Seiler.

8. John N. Gamewell's testimony, Gamewell & Co. v. Allen.

9. Testimony on behalf of Gamewell and Company, Gamewell & Co. v. Allen. When the company later established its own manufactory, these shops became the center of continued innovation (see testimony in Ruddick v. Milliken).

10. Gamewell Fire-Alarm Telegraph Company, *Fire Alarm Telegraphs*, 7–8; Greer, *Alarm Security*, 22–23; Gamewell v. Chester; Gamewell v. Chester and Chester; Gamewell v. New York; Gamewell v. Pearce and Jones; Gamewell Fire-Alarm Telegraph Company v. City of Chillicothe, *Federal Reporter* 7:351; Gamewell Fire-Alarm Telegraph Company v. Municipal Signal Company, *Federal Reporter* 52:471, 77:490; Municipal Signal Company v. Gamewell Fire-Alarm Telegraph Company, *Federal Reporter* 52:459; Gamewell Fire-Alarm Telegraph Company v. Mayor and City Council of City of Bayonne, N.J., *Federal Reporter* 194:146.

11. Agreements between Calahan and Martin V. B. Finch, 5 April 1867, and among Elisha Andrews, George Conant, and Gold and Stock, 30 November 1867, EC; certificate of incorporation of the Gold and Stock Telegraph Company, Gold and Stock v. Pearce and Jones; "The Gold and Stock Telegraph Company," *Telegrapher* 6 (1868–69): 12; Reid, *Telegraph in America*, 602–13; terms of agreement between Gold and Stock and Samuel Laws, G&SM (1867–70), 57–60.

12. G&SM (1867–70), 141–45.

13. Reid, *Telegraph in America*, 621–22; "The Financial and Commercial Telegraph," *Telegrapher* 6 (1869–70): 124; TAEB 1, Docs. 97, 130, 164; G&SM (1867–70), 57–60, 65–66, 73–76, 115, 118, 122–23, 126–30, 135, 141–49 157–63; (1870–79), 5, 13–14.

14. TAEB 1, Docs. 91, 92, 131, 136, 164–65, 173, 195, 211, 254, 262, and chap. 10 introduction.

15. Lefferts also secured exclusive leases with both the Gold and the Stock exchanges. Reid, *Telegraph in America*, 621–22; "The Financial and Commercial Telegraph" (see n. 13 above); G&SM (1867–70), 23–28, 73–76, 97, 122–23, 126–30, 141–49, 162–63; (1870–79), 3, 7.

16. Aware of Western Union's interest in this field, Lefferts initiated the negotiations in the hope that he could prevent competition from the giant company in Gold and Stock's main market in New York by offering to cooperate with the industry giant in other cities. Originally the two men considered establishing a separate company to develop business outside of New York City, which was reserved to Gold and Stock. Orton, however, expressed concern over the time and money required to develop a new company for this purpose and proposed that Gold and Stock carry on the business itself under Western Union's control. After extensive negotiation, this was agreed to, with Gold and Stock acquiring control of Western Union's patents as well as its commercial news department. Although Lefferts wished to retain a degree of autonomy for his company, Western Union's ability to threaten its New York market if competition ensued probably convinced him to agree to Orton's offer. Orton to Anson Stager, 12 and 15 November 1870, 10 and 24 January and 24 February 1871, LBO 8:161–70, 173–80, 317–21, 428–39; Orton to

Stager, 22 March 1871, LBO 9:35–39; WUEM A:435, 479; G&SM (1870–79), 3–5, 27–28, 44, 55–59, 64–78.

17. Orton to Frank Ives Scudamore, 11 August 1873, LBO 12:230–32.

18. *TAEB* 1, Doc. 164; G&SM (1867–70), 49–51, 56–59, 64–77, 122–23, 126–30, 142, 154–55, 159–62; (1870–79) 149–51, 167, 182; G&SEM (1871–78), 9–10, 17–74; (1871–78) 68–74; MM; RL.

19. Healy, "Stock Tickers," 120–21; testimony and exhibits in Wiley v. Field; Gold and Stock v. Commercial; Commercial v. Gold and Stock; affidavit of Alfred S. Brown, Western Union v. B&O; RL.

20. Reese Jenkins develops the notion of a business-technology mindset in his *Images and Enterprise.*

21. Reid, *Telegraph in America,* 520–27; "William Orton," *Journal of the Telegraph* 11 (1878): 129–30; *Dictionary of American Biography,* s.v. "Orton, William."

22. William Orton to George F. Davis, 8 January 1865, LBO 1:57–58.

23. LBO 5:50.

24. Orton to Division Superintendents, William Eckert, Anson Stager, and John Van Horne, 7 January 1869, LBO 5:171.

25. Western Union, *Statement,* 79.

26. Orton to George B. Prescott, 19 November 1868, LBO 4:471.

27. Jenkins, *Images and Enterprise,* 33, and Lubar, "Corporate and Urban Contexts," 194–201.

28. For Varley's investigations see Varley, "Report"; Orton to Varley, 23 November 1867, LBO 2:304; "Changing the Alphabet," *Journal of the Telegraph* (2 December 1867): 4; "Mr. Varley's Report," ibid. (16 March 1868): 4; "Effects of Cold on Insulation," ibid. (1 April 1868): 4; Reid, *Telegraph in America* (1886), 534–35.

29. For example, when Charles S. Jones of the Western Union Albany office offered the company a new switchboard patent for $500, Orton asked William Hunter, the superintendent of supplies to determine if it was a significant improvement on those currently being used by the company. On another occasion Orton asked George M. Phelps, superintendent of the company's factory and an inventor of important improvements in printing telegraphs, to examine a printing telegraph devised by Edwin F. Ludwig of the company's commercial news department. Western Union may have paid the cost of building two experimental instruments for Ludwig. Orton to C. S. Jones, 29 May 1868, LBO 3:464; Orton to William Hunter, 17 June 1868, LBO 4:39; and Orton to Edwin F. Ludwig, 6 February 1869, LBO 5:225.

30. On the Western Union investigation of insulators see letters from Orton to Anson Stager, David Brooks, T.B.A. David, from March, April, and May 1868 in LBO 3.

31. Reid, *Telegraph in America,* 530–36, 563–66, 674–79, and Orton to Division Superintendents, 23 October 1870, LBO 8:92–93.

32. Orton to Anson Stager, 23 November 1870, LBO 8:193–200.

33. "The Late President Orton," *The Operator* (1 May 1878): 6.

34. It has been estimated that business use of the telegraph constituted about 80 percent of total traffic and that most of the press usage was for commercial and financial news. As late as 1887, the president of Western Union could claim "that not more than two per cent of the entire populations ever use the telegraph in one year, and not over five per cent of the revenues of the telegraph is derived from family and social news." Marshall Lefferts had reflected upon this modest use by individuals thirty years earlier, noting that "citizens, as a general thing, have no conception of the amount of business daily transacted over the wires." Both Green and Lefferts pointed to business transactions as the principal source of revenue for the industry, with newspaper copy as an important

but distant second. Norvin Green to Postmaster General William Vilas, 17 November 1887, LBO 8:427–33; Lefferts, "Electric Telegraph," 251; DuBoff, "Business Demand."

35. Western Union, *Proposed Union;* Kieve, *Electric Telegraph,* 183–85, 191–96, 216–22, 230; Butrica, "From *Inspecteur* to *Ingénieur*"; Attali and Stourdze, "Birth of the Telephone," 97–102.

36. Western Union, *Annual Report* (1869), 39.

37. Orton to James D. Reid, 21 September 1869, LBO 6:405–10.

38. Western Union also used Phelps's "combination printer" to a very limited extent on its lines, primarily between New York and Boston (Fischer and Preece, "Joint Report," 383–85). Only in France, where the Hughes printing telegraph was extensively improved and used on the main lines of the government telegraph, did printing telegraphs play an important role on intercity telegraph lines before the end of the century. On the French development of the Hughes printing telegraph system see Butrica, "From *Inspecteur* to *Ingénieur,*" chap. 3.

39. Craig, *Machine Telegraph of Today* and his numerous letters and articles in *The Telegrapher* and *TAEB* between 1870 and 1872; Reid, *Telegraph in America* (1886), 778–80; Patrick B. Delany's testimony, Harrington and Edison v. A&P; S. Doc. 291.

40. The advantages of the Morse key and sounder system led the British Post Office Telegraph Department to adopt it for most ordinary messages in the late 1870s. Previously, the department had relied largely on the Wheatstone automatic. Preece, *Recent Advances,* 20–21.

41. Orton to R. Smith, Editor, Cincinnati *Gazette,* 15 July 1869, LBO 6:225–35; Orton to James D. Reid, 21 September 1869, LBO 6:405–10; Orton to Anson Stager, 21 March 1872, LBO 9:465–69; Orton to George Little, 27 October 1874, LBO 13:413; Orton to Daniel H. Craig, 19 February 1876, LBO 16:258; Western Union, *Annual Report* (1869), 37–39; (1874): 15–16; (1875): 22–23; "Mr. Craig's Reply to Mr. Orton's Challenge," *Telegrapher* 7 (1870–71): 25–26; and "The Postmaster-General and Mr. Orton on Automatic or Fast Telegraphy," ibid. (1874): 10–11.

42. The Franklin Company began operation in 1866 with only two wires between Boston and New York and added four other wires between Boston and Washington after combining with the Insulated Lines Company the following year. Western Union, *Annual Report* (1869), 34.

43. James G. Smith's testimony, Western Union v. B&O. Soon after the introduction of the Stearns duplex on its lines, the Franklin and another competing company, the Atlantic and Pacific Telegraph Company, considered consolidating their lines. During the course of their negotiations, in the winter of 1869–70, Stearns tested his instrument on the Atlantic and Pacific's long line between New York and Buffalo. The Atlantic and Pacific also tested a double transmitter (which was either a duplex or a diplex to transmit two messages in a single direction) devised by Thomas Edison, whose experiments were supported by E. Baker Welch, a director of the Franklin Company. The failure of both these instruments and a growing disagreement between the two companies prevented consolidation and further development of duplex telegraphy by either company. *TAEB* 1, Docs. 61, 63, 68.

44. For the best account of the Congressional grant of the patent to Page see Post, *Physics, Patents, and Politics,* 173–81. Post also cites several articles from telegraph and electrical journals which give an indication of the concern felt within the telegraph industry. For Western Union's side of the story see Orton to Cambridge Livingston, 22 December 1868, LBO 5:114–16; Orton to Anson Stager, 3 February 1871, LBO 8:367–73; Orton to Sen. O. S. Terry, 17 February 1875, LBO 14:197–203; and affidavit of Norvin Green, Western Union v. Tillotson.

45. Lester G. Lindley analyzes this debate in *The Constitution Faces Technology*.

46. Hubbard also pointed to the adoption of Charles Wheatstone's automatic telegraphy by the newly nationalized British telegraph. Western Union, *Statement*, 78–79.

47. George Prescott, to R. S. Culley, 12 October 1871, 9 November 1871, 5 and 15 February 1872, "Defendant's Exhibits" in Western Union v. B&O.

48. The company also acquired the rights to Stearns's invention for the Dominion of Canada. Orton to Nathan C. Ely, 17 November 1872, LBO 11:132–34; Orton to John Van Horne, 14 May 1873, LBO 11:451–53; Orton to James Dakers, 22 September 1873, LBO 12:307–11; and Orton to Joseph B. Stearns, 24 September 1873, LBO 12:320–23; WUDM A, 371–73; WUEM B, 28, 69.

49. Orton to Joseph B. Stearns, 2 December 1874, LBO 14:34–38. Also see Orton's testimony in Atlantic and Pacific v. Prescott and Others, *TAEM* 10:58–154.

50. Testimony of Thomas A. Edison, Atlantic and Pacific v. Prescott and Others, *TAEM* 9:496.

51. *TAEB* 2, Docs. 694–95.

52. Orton to George M. Phelps, 8 September 1875, LBO 14:462; WUEM B, 433; WUEM C, 288; Elisha Gray testimony, "Telephone Interferences." Also see Hounshell, "Elisha Gray and the Telephone," 133–61.

53. Bruce, *Bell*; and Hounshell, "Elisha Gray and the Telephone," 133–61.

54. Edison, to Orton, c. March 1877, *TAEM* 14:226–27; draft agreements between Edison and Western Union, c. March 1877, *TAEM* 14:228–35; and agreement between Edison and Western Union, 22 March 1877, *TAEM* 28:1029–31.

55. WUEM C, 10–11, 23–24, 28.

56. Green to Gardiner Hubbard, 1 December 1879, LBG 2:456–62.

57. Green to Gen. Wager Swayne, 29 March 1881, LBG 3:393–97.

58. It was probably during these negotiations that Bell's father-in-law and business associate Gardiner Hubbard made the famous offer to sell Bell's patents on the telephone to Western Union for $100,000. Norvin Green's testimony, *Capital and Labor*, 882.

59. Orton expressed much greater willingness to pursue the competition between the firms than did Green, but he may have been motivated in this case by personal animosity toward Gardiner Hubbard, president of American Bell. It is probable, however, that he would have agreed with the outcome of Green's negotiations. Norvin Green to Charles A. Cheever, 12 February 1878, LBG 1:365–71; Green to Samuel S. White, c. 8 May 1878, LBG 2:134–38; Green to George Gifford, 8 and 19 July and 18 August 1879, LBG 2:233–35, 275–80, 314–17; Green to William H. Forbes, 3 September 1879, LBG 2:335–41; agreement between Western Union, American Speaking Telephone Company, Gold and Stock Telegraph Company, and American Bell Telephone Company, 10 November 1879, AT&T.

60. Reich, "Struggle to Control Radio," 233–34.

61. Green to George B. Prescott, 28 March 1879, LBG 2:72–74.

62. WUEM D, 150; Gerritt Smith's testimony, Western Union v. B&O.

63. For Field's work see patent interferences Field v. Pope and Field v. Lugo; "Dynamo-Electric Machines Substituted for Batteries in Telegraphy," *Journal of the Telegraph* 13 (1880): 33; and Maver, *American Telegraphy*, 217–25. For Brooks see Reid, *Telegraph in America* (1886): 748–52; Prescott, *Electricity and the Electric Telegraph* (1885), 1076–78; Brooks Underground Telegraph Company, Certificate of Incorporation, Charter Book B, WUS; "Brooks Insulated Underground Cable, *Journal of the Telegraph* 12 (1879): 149; and "Underground Telegraph in Cities," ibid., 166.

64. Subsequent improvements in the company's technology were largely the work of the company's electrical department. Reid, *Telegraph in America* (1886), 674–79, 731,

735–36; Finn, "Wheatstone Automatic System," 468–71 (this article was the last of a series by Finn on the Wheatstone automatic). For a study of technological change in a mature industry see Usselman, "Running the Machine."

65. Green to G. Kyle, 4 October 1882, LBG 4:334–37. Green raised questions concerning Western Union's support of Edison's inventions in letters to the inventor of 23 December 1878 and 12 March 1880, Box 8, WUS.

66. For the Automatic Telegraph Company see *TAEB,* vols. 1 and 2. For the history of American Rapid see Reid, *Telegraph in America* (1886), 778–81; "The American Rapid Telegraph Company," *Operator* 10 (15 October 1879): 4–5; advertising pamphlets of the American Rapid Telegraph Company, Box 1, WUS; Boston Safe v. American Rapid; Boston Safe v. Bankers and Merchants; Harland v. American Rapid; Robeson v. American Rapid; and the following material in Patent Assignment Digests, U.S. Patent Office Records, DNA: Agreement between Bear and Angle, 17 April 1878, Vol. B-11; Memorandum of Agreement between Foote, Randall, and Craig, 1 July 1874, and Agreement between Foote, Randall, and Craig, 28 December 1875, Vol. F-4; Agreement between Foote, Randall, Craig, Bear, and Angle, 11 December 1878, Vol. F-5.

67. Customers could deposit letters in special post office boxes for transmission during the night when few ordinary messages were being transmitted. These were to be sent at significantly lower costs. In 1875, Edison had proposed that Atlantic and Pacific use his automatic telegraph to provide a similar service (*TAEB* 2, Docs. 590, 596, 598). In the 1890s, Patrick Delany's automatic telegraph system was used by the Telepost Company for night letters (see n. 69 below).

68. Green, like Orton, believed that its business traffic was the mainstay of the company and held similar opinions on the failure of automatic telegraphy to meet the requirements of such service. Unlike Orton, however, he felt automatic telegraphs could play an important role in press service. Green to Daniel H. Craig, 20 May 1878, LBG 21:483–89; Green to Gardiner Hubbard, 1 December 1879, LBG 2:456–62; and Thomas T. Eckert to Norvin Green, 18 January 1882, WUEM F, 140–42.

69. Patrick Delany, a telegraph inventor at one time connected with the Automatic Telegraph Company, developed another system of automatic telegraphy in the 1890s which also failed to compete successfully with Western Union. This instrument was later used about 1910 for a letter service known as Telepost. S. Doc. 291; advertising brochure of the Telepost Company, *TAEM* 9:270–79; Robert C. Clowry's testimony, *Report of the Joint Committee,* 656–57.

70. Reid, *Telegraph in America,* 580–94; testimony and exhibits, Harrington and Edison v. A&P.

71. Testimony and exhibits, Harrington and Edison v. A&P; Fischer and Preece, "Joint Report," 386–90.

72. Gould's claims to the quadruplex were based on Edison's agreements with the president of Automatic Telegraph, George Harrington, and led to a protracted legal battle with Western Union. The litigation over Edison's quadruplex involved a series of suits and countersuits that became known as the "Quadruplex Case." The several volumes making up the printed record of the Quadruplex Case are found in *TAEM,* Reels 9–10; also see *TAEB* 2. Klein (*Jay Gould,* 201), argues that Gould's primary interest in acquiring Edison's inventions, as well as rights to the Wheatstone automatic, stemmed from his desire to upgrade Atlantic and Pacific's telegraph system. However, the quadruplex was never used on the company's lines and after providing some initial funds, Gould failed to support Edison's work on automatic telegraph improvements.

73. The validity of the D'Infreville patents was not challenged after Atlantic and Pacific and the patent came under the control of Western Union (affidavit of Thomas T. Eckert,

Western Union v. B&O, and Reid, *Telegraph in America* [1879], 590). The Atlantic and Pacific also provided funds for D'Infreville to conduct some of the earliest experiments with dynamos for use in telegraphy (Georges D'Infreville's testimony, Von Hefner Alteneck v. Western Union).

74. Reid, *Telegraph in America,* 586 (1879), and 579 (1886); Western Union, *Annual Report* (1877), 11.

75. Lindley, *The Constitution Faces Technology,* 243–46.

76. Reid, *Telegraph in America* (1886), 577–80.

77. Affidavits of Thomas T. Eckert and Henry C. Townsend, Western Union v. B&O; Western Union's bill of complaint, Western Union v. American Union; Henry Van Hoevenbergh's testimony, Cochrane v. Van Hoevenbergh.

78. Reid, *Telegraph in America* (1886), 580–82.

79. Czitrom, *Morse to McLuhan,* 24–29; Lindley, *The Constitution Faces Technology,* 248–49. Lindley's book focuses on the earlier wave of antimonopoly sentiment faced by Western Union following the Civil War.

80. Much of the debate occurred in Congressional hearings concerning proposals to establish a postal telegraph service and in other government reports on the telegraph industry. See S. Repts. 577, 805; S. Doc. 291; *Labor and Capital;* U.S. Postmaster General, *Argument;* Parson, *Telegraph Monopoly;* and U.S. Industrial Commission, *Report on Transportation.*

81. Reid, *Telegraph in America* (1886), 748–52; Brooks Underground Telegraph Company, Certificate of Incorporation, Charter Book B, WUS; N.Y. S. Doc. 82; "Brooks Insulated Underground Cable," *Journal of the Telegraph* 12 (16 May 1879): 149; "Underground Telegraph in Cities," ibid. (1 June 1879): 166; and the following pages in Green's letterbooks: LBG 3:367, 486; 4:62–69, 71–73, 74–79, 161–68, 221–24; 5:20, 46–52, 6:59–72, 409–14; and 7:88–97, 145–51. For Orton's earlier response to this issue see Western Union, *Review of the "Opinion of Experts."*

82. Reid, *Telegraph in America* (1886), 758–60. For Bankers and Merchants, which was organized in 1881 by some Wall Street financiers unhappy with the consolidation between Western Union and American Union see ibid, 782–87; Boston Safe v. Bankers and Merchants; and Harland v. American Rapid.

83. Defendant's Exhibits, Affidavit of John Van Horne and Pope Opinions 1–3; and Baltimore and Ohio Telegraph Company, "Correspondence and Opinions Relating to Validity of Quadruplex and Condenser Patents," New York Public Library; and Pope Report on Quadruplex Telegraphy and Patents, WUC, all part of Western Union v. B&O.

84. Decision of Judge Wallace on Motion for Preliminary Injunction, Western Union v. B&O, *Federal Reporter* 25:36.

85. After the consolidation, Western Union was able to obtain a permanent injunction.

86. Postal Telegraph and Cable had been created through the consolidation of the Bankers and Merchants and the Postal Telegraph Company, established in 1881 on the basis of Gray's harmonic telegraph system and a facsimile telegraph developed by W. A. Leggo, with the Commercial Cable Company by John W. Mackay, the "Silver King" of Comstock fame. Reid, *Telegraph in America* (1886), 772–77; Harlow, *Old Wires and New Waves,* 424–28; "New Telegraph Lines," *Telegrapher* 12 (1881): 300; *Operator* 12 (1881): 285; and testimony of Clarence Mackay and Edward J. Nally, *Joint Committee,* 685–87, 726.

87. Taltavall, *Telegraphers of To-Day,* 162; *Report of the Joint Committee,* 7–8, 581–628.

88. Letwin, *Law and Economic Policy.*

89. Prindle, *Patents as a Factor in Manufacturing,* 14–16.

Chapter 6. From Shop Invention to Industrial Research

1. Hughes, "Evolution of Large Technical Systems," 76–80; Hughes, *Networks of Power*, chap. 6; Hughes, "Technological Momentum in History."

2. These engineering values and the corporate bureaucracies that helped spawn them have been the subject of several important studies. Among the best are Calvert, *Mechanical Engineer in America;* Layton, *Revolt of the Engineers;* McMahon, *Making of a Profession;* and Sinclair, *Centennial History.*

3. Ferguson, "Mind's Eye"; Layton, "American Ideologies"; Layton, "Mirror-Image Twins."

4. Taltavall, *Telegraphers of To-day,* 165–67; Reid, *Telegraph in America* (1886), 789; Maver, *American Telegraphy,* 226; Francis W. Jones's testimony, Buckingham v. Jones.

5. Taltavall, *Telegraphers of To-day,* 255; Reid, *Telegraph in America* (1886), 675–76.

6. *Report of the Joint Committee,* 581, 609, 652–57, 693; S. Doc. 291.

7. S. Doc. 291; Houston, "Facsimile Telegraph"; advertising brochure of the Telepost Company, *TAEM*-9:270–79; Delany to Thomas Eckert, 15 October 1894 and 25 January 1895, Supports, Cable—Inventions and Patents File, Western Union Engineering Library.

8. The Delany multiplex had significant problems on long-distance lines and was used only experimentally in the United States. The system was apparently based on the "phonic wheel" developed by Danish inventor Paul LaCour in 1878, U.S. patent rights of which were purchased by the Standard Multiplex Telegraph Company. Delany took out several subsequent patents, including one with Edward Calahan, and developed a system that became known as the Delany multiplex. It appears that LaCour probably invented the time-sharing system, but that Delany devised a more effective means for synchronizing the instruments, as evidenced by the Franklin Institute's award of the Elliot Cresson Medal to Delany for his work on synchronism, and of the John Scotte Legacy Medal to LaCour for his phonic wheel. U.S. Patents 203,423, 281,339, 286,273–78, 286,281–82; pamphlet of the Standard Multiplex Telegraph Company, c. 1884, Engineering Societies Library, New York; Houston, "Synchronism"; Houston, "Delany Synchronous Multi-Plex"; Maver, *American Telegraphy,* 343; Maver and McNicol, "American Telegraph Engineering," 1312; letter to the author from Jytte Thorndall, curator, Danish Museum of Electricity, 12 October 1990.

9. Delany, "Telegraph Line Adjustment," 150. For general biographical information on Delany see Taltavall, *Telegraphers of To-Day,* 243–44.

10. Edison complained to Gould that Eckert required him "to wait for days and weeks for small sums due, doubting my honesty in the transaction, of four dollars; urging me to produce things and then refuse to pay for them—granting no money to conduct experiments, and then say I am not doing anything for the Co" (Edison to Jay Gould, 2 February 1877, Complainant's Exhibit 4, Apl. 11, '05, Harrington and Edison v. A&P). For Edison's role as Atlantic and Pacific electrician see *TAEB* 2, chaps. 7–9, and Docs. 522, 648, 672.

11. On Elisha Gray see *Dictionary of American Biography,* s.v. "Gray, Elisha"; Dixon, "Telautograph"; and material on his telautograph in folder 11 of GP.

12. "Extract from a Lecture Delivered by W. H. Preece," *Journal of the Telegraph* 18 (1885): 18.

13. Estimates placed "broker" business at 40 percent of the total volume of telegraph traffic. Delany, "Rapid Telegraphy," 3:464–65; discussion of Potts, "Rowland Telegraphic System," 539–46; discussion of Vansize, "New Page-Printing Telegraph," 37–43.

14. For the origins of telegraph company officials see Taltavall, *Telegraphers of To-Day.*

Discussion of operators moving into management is found in *Labor and Capital*, 3:886–91; U.S. Industrial Commission, *Report on Transportation*, 224; "Governor Cornell's Lecture," *Telegraph Age* 14 (1894): 48; "The Qualifications of Good Managers," ibid., 97; also see chap. 5, nn. 31–33 above.

15. Discussion of Vansize, "New Page-Printing Telegraph," 40–43; discussion of Crehore and Squier, "Synchronograph," 143–44; discussion of Potts, "Rowland Telegraphic System"; discussion of Maver and McNicol, "American Telegraph Engineering," 1350–54; *Report of the Joint Committee*, 653–57, 694–95, 845–50.

16. In the late 1870s British telegraph officials, with their greater commitment to machine technology, adopted Morse technology for most regular messages because they agreed that it reduced errors. Even in France, where government messages played a large role, and machine systems, notably printing telegraphs, were used as the principal system on main lines, the Morse found extensive service for much ordinary traffic on less heavily trafficked lines. Israel and Nier, "Transfer of Telegraph Technologies," 104–9; Preece, *Recent Advances*, 20–21; Kieve, *Electric Telegraph*, 234–35; Butrica, "From *Inspecteur* to *Ingénieur*," 59–60.

17. On the declining status of telegraph operators and the role of female operators see Gabler, *American Telegrapher*, chaps. 3–4.

18. Bliss, "Telegraph Operator," 242.

19. "The Condition of Telegraph Operators," *Journal of the Telegraph* 11 (1878): 244.

20. On technical changes see Maver, *American Telegraphy*; McNicol, *American Telegraph Practice*; and Maver and McNicol, "American Telegraph Engineering." On the quadruplex as a source of advancement see "Philadelphia Postal Notes," *Telegraph Age* 14 (1894): 72.

21. Gartman, *Auto Slavery*. Gartman was responding to Harry Braverman's arguments in *Labor and Monopoly Capital*.

22. Delany, "Telegraph Line Adjustment," 507–8.

23. Fischer and Preece, "Joint Report," 588–90.

24. On company policies that resulted in speeding up see U.S. Commission on Industrial Relations reprinted in the *Commercial Telegraphers Journal* 13 (1915): 181–82, 301, 325–26. On the typewriter and the Martin mecograph, see ibid., esp. 157, 182, 315; letter to the editor, 27 January 1873, *Telegrapher* 9 (1873): 38; Maver, *American Telegraphy*, 74–77; Maver and McNicol, "American Telegraph Engineering," 1316–19; Vansize, "New Page-Printing Telegraph," 9–10; and Barclay, "High Speed Printing," 3:505–6.

25. Gabler, *American Telegrapher*, chaps. 1, 5; Ulriksson, *Telegraphers*, chaps. 4–5; *Commercial Telegraphers Journal* 13 (1915): 145–328. Because it was less physically taxing, the mecograph key may also have led managers to begin using women more extensively on heavily trafficked lines even before the strike.

26. Kinsley, "High-Speed Printing Telegraph," 1243; *Joint Committee*, 657–58, 693–94.

27. Barclay, "High-Speed Printing"; Potts, "Printing Telegraphy"; Potts, "Rowland Telegraphic System"; Kinsley, "High-Speed Printing"; Vansize, "New Page-Printing Telegraph"; Maver, *American Telegraphy and Encyclopedia*, 436a–36x; Maver and McNicol, "American Telegraph Engineering," 1340; McNicol, *American Telegraph Practice*, 421–23, 471–72; Hausmann, *Telegraph Engineering*, 121–34; Bell, "Printing Telegraph Systems"; Reiber, "Printing Telegraph Systems"; U.S. Bureau of the Census, *Special Reports*, 200–202.

28. Davies, *Women's Place*.

29. Some information on the introduction of the new technology is found in *Commercial Telegrapher's Journal* 13 (1915): 153, 257, 266–73.

30. Fischer and Preece, "Joint Report," 341–45. This did not mean that scientific

theory and higher physics were an integral part of telegraph engineering in Great Britain, but rather that telegraph engineers there were attempting to develop an engineering science (Hunt, "Practice vs. Theory").

31. Wasserman, *From Invention to Innovation,* 24–30. The growing use of cables in French telegraphy and the development of high-speed telegraph systems produced a similar movement toward engineering science (Butrica, "From *Inspecteur* to *Ingénieur*," 116–17, 235–50).

32. Laboratory notebook, Cat. 287:12(2), *TAEM* 5:478.

33. U.S. Pat. 146,311 (10 February 1873).

34. Jones, "Static Difficulties in Telegraph Wires," 120–27.

35. Discussion of Fowle, "Telegraph Transmission," 1739. It should be noted that there were a few papers that discussed both telephone and telegraph transmission, but these were all by telephone engineers.

36. McNicol, *American Telegraph Practice,* v, and Hausmann, *Telegraph Engineering.*

37. Discussion of Rhoads, "Railroad Telegraph and Telephone Engineering," 417.

38. Ibid., 350–79.

39. Maver's discussion in ibid., 399. Also see Maver's description of one graphic method in Maver and McNicol, "American Telegraph Engineering," 1331, and his use of another in Maver, *American Telegraphy,* 122–24, 204–6, 220–23.

40. Discussion of Rhoads, "Railroad Telegraph and Engineering," 408–10.

41. For a discussion of engineering design and engineering science see articles cited in n. 3 above.

42. Friedel and Israel, *Edison's Electric Light,* esp. chaps. 2–3; Hughes, *Networks of Power,* chap. 7; McMahon, *Making of a Profession,* chap. 2; Millard, *Business of Innovation,* esp. chap. 2; R. Rosenberg, "American Physics"; R. Rosenberg, "Test Men, Experts"; Wasserman, *From Invention to Innovation,* chaps. 2–4.

43. "The Probability of Improvement in Our Telegraphic System," *Telegrapher* 5 (1869): 414.

44. "Activity of Electrical Research and Telegraphic Invention," *Journal of the Telegraph* 13 (1880): 212.

45. On lectures for telegraphers see *Telegrapher* 2 (1866–67): 84, 98, 110, 150; letters, articles and editorial concerning the establishment and meeting of various electrical societies formed by telegraphers appeared in *Telegrapher* 7 (1870–71): 53; 10 (1874): 231; 11 (1875): 32, 359; and *Journal of the Telegraph* (1878): 36–38, 50, 82–83; (1879): 38, 40, 358; (1880): 200, 293, (1881): 21–23, 30, 68; On the American Electrical Society also see *Journal of American Electrical Society* 1–3 (1875–80). On the New York Electrical Society also see the Society's *Constitution and By-Laws* (1881). On the origins of the American Institute of Electrical Engineers see McMahon, *Making of a Profession,* chap. 1; McMahon, "Corporate Technology," 1383–90; and R. Rosenberg, "Test Men, Experts," 207.

46. McMahon, *Making of a Profession,* 36–40.

47. Edison to William Orton, c. January 1877, document file folder 77-002, *TAEM* 14:226. On Edison and *Science* see document file folders 80-041, 81-044, 82-051, *TAEM* 55:360–492, 59:451–608, 63:661–71.

48. Hounshell, "Edison and the Pure Science Ideal," touches upon these contrasting views of science. W. Bernard Carlson discusses the role of craft knowledge in science in "Invention, Science, and Business."

49. On De Forest see Aitken, *Continuous Wave,* 542–48. On Kettering see Leslie, *Boss Kettering.* On engineers and scientists in corporate bureaucracies see Layton, *Revolt of the*

Engineers; Reich, *Making of American Industrial Research;* Wise, "New Role for Professional Scientists"; Wise, *Willis R. Whitney;* and Noble, *America by Design.*

50. On Western Electric see G. D. Smith, *Anatomy of a Business Strategy;* Wasserman, *From Invention to Innovation,* 107–10; and Reich, *Making of American Industrial Research,* 159–60.

51. Fagen, *History of Engineering and Science,* 732–58, and Bell, "Printing Telegraph Systems."

52. More typical were the views of General Electric's managers as described in Reich, *Making of American Industrial Research,* 60–61, and Carlson, "Myth of the Heroic Inventor."

53. Western Union, "Brief Outline of Technical Progess," 28.

54. In his letter to Orton, Edison also stated that "at present the cost of running the machine Shop including Coal Kerosene & Labor is about 15 per day or 100 per week." Edison to Orton, c. January 1877, document file folder 77-002, *TAEM* 14:226.

55. Israel, "Telegraphy and Edison's Invention Factory", and Friedel and Israel, *Edison's Electric Light,* chaps. 1–2.

56. John Irwin, an independent inventor who worked on a variety of inventions, including telephony, also had a small laboratory and machine shop at his home. Testimony of John Irwin and William Lienhardt, "Telephone Interferences," 60–63, 78, 94–96.

57. Wasserman, *From Invention to Innovation,* and Reich, *Making of American Industrial Research,* 137, 142–44.

58. Maver and McNicol, "American Telegraph Engineering," 1356.

59. Leahey, "Skilled Labor"; Schatz, *Electrical Workers,* chap. 2. On scientific management see Nelson, *Frederick W. Taylor;* Nelson, *Managers and Workers;* Aitken, *Scientific Management in Action;* Calvert, *Mechanical Engineer in America;* Layton, *Revolt of the Engineers;* Sinclair, *Centennial History;* and Noble, *America by Design.*

60. David S. Landes describes their role in the development of modern clock technology in "Hand and Mind in Time Measurement."

61. U.S. Bureau of the Census, *Census Reports Volume X,* 186.

62. "J. H. Bunnell & Co.—A Magic Name in Telegraphy," *Dots and Dashes* (October-November 1986): 14 (reprinted from the *Railroad Telegrapher,* March 1950). Although J. H. Bunnell and Company continues to exist as a specialized electrical manufacturer, the current owners would not allow me to look at the company's old records.

63. Gamewell Fire-Alarm Telegraph Company, *Fire Alarm Telegraphs,* 8; testimony and exhibits, Page Machine v. Dow Jones.

64. Western Union, "Brief Outline of Technical Progress," 28–29.

65. Maver and McNicol, "American Telegraph Engineering," 1356.

66. Western Union, "Brief Outline of Technical Progress," 7, 13–16, 29–30.

67. In discussing the impact of scientific management labor historian David Montgomery notes that it "was as important an element in the transformation of American ideology as it was in the reshaping of industry" (*Fall of the House of Labor,* 249–50). For other discussions of scientific rationalization and corporate and social reform see Akin, *Technocracy and the American Dream;* Haber, *Efficiency and Uplift;* Hays, *Conservation and the Gospel of Efficiency;* Layton, *Revolt of the Engineers;* Sklar, *Corporate Reconstruction;* Weinstein, *Corporate Ideal;* and Wiebe, *Search for Order.* There is also a growing literature dealing with the 1920s, including articles appearing in Hawley, *Herbert Hoover,* and in *Business History Review* 52 (Autumn 1978).

Afterword

1. In contrast the Naval Consulting Board headed by Thomas Edison represented an unsuccessful attempt to mobilize the nation's independent inventors. The results of the Board's work were disappointing; only about one hundred of the more than one hundred thousand suggested inventions merited further investigation, and only one device was actually built. Noble, *America by Design,* 148–66; Kevles, *Physicists,* chaps. 8–9; Dupree, *Science in the Federal Government,* 306–15; Hughes, *American Genesis,* 118–37.

2. The National Research Council's survey found over sixteen hundred laboratories by 1931. National Research Council, *Research Laboratories.*

3. Kaempffert, "Invention as a Social Manifestation," 64. Kaempffert was the editor of a two-volume *Popular History of American Invention* and author of *Invention and Society.* Curiously, Kaempffert reported in this latter work that Prof. Hornell Hart had studied the 171 inventors mentioned in the *Popular History* and concluded that most had attended college or taken technical courses, while only one third were self-taught. Furthermore, most inventors worked in fields related to their inventive activity and that many more achieved some degree of financial success than was commonly believed. Nonetheless, the power of popular perceptions of nineteenth-century inventors was such that in his 1932 essay, Kaempffert could still portray them as lone struggling geniuses.

4. Most popular works on invention implicitly accept the heroic view by treating the history of invention as a catalogue of great inventors and inventions. This is explicit in the title of L. Sprague de Camp's *The Heroic Age of Invention* (1961). For other examples see Ernest Adams, *Mechanical and Electrical Inventions* (1900); Frank Bachman, *Great Inventors and Their Inventions* (1918); Roy Baker, *The Boy's Book of Invention* (1899) and *The Boy's Second Book of Invention* (1903); Roger Burlingame, *March of the Iron Men* (1946); E. W. Byrn, *The Progress of Invention in the Nineteenth Century* (1900); Russell Doubleday, *Stories of Inventors* (1904); Rupert S. Holland, *Historic Inventions* (1911); George Iles, *Leading American Inventors* (1912); William Mowry, *American Inventions and Inventors* (1900); Robert Routledge, *Discoveries and Inventions of the Nineteenth Century* (1903); and Holland Thompson, *The Age of Invention* (1921). There are also a host of biographies of famous inventors that attribute to them traditional heroic qualities.

5. A general survey of scholarly research on the development of industrial research is found in J. K. Smith, "Scientific Tradition." Also see Dennis, "Accounting for Research." No comparable literature exists for other types of inventive activity, but see Whalley, "Independent Inventing" and von Hippel, *Sources of Innovation.*

6. Reich, *Making of American Industrial Research,* 3.

7. Contemporary accounts of industrial research are Fleming, *Industrial Research in the United States,* and the National Research Council, *Research Laboratories* (1920–21, 1927, 1933). The listings in the National Research Council's surveys of research laboratories indicate that many laboratories employed a large number of engineers and undertook much engineering research.

8. There is a need for more thorough historical investigations of Western Electric's role in the development of twentieth-century telephone technology. Some hints of its role can be found in Bode, *Synergy,* and Bell Telephone Laboratories, *Engineering and Operations,* 19–24.

9. For one important account of this change see R. Rosenberg, "Academic Physics."

10. Calvert, *Mechanical Engineer in America.*

11. Layton, "Mirror-Image Twins."

12. See, for example, Reich, "Irving Langmuir," and Wise, "New Role for Professional Scientists." In academic science instrumentation has also taken on increasing importance.

13. Kranzberg, "U.S. Engineering Curriculum."

14. See Kline, "Science and Engineering Theory."

15. Kidder, *Soul of a New Machine,* 95–96, 162–63, 192, 195, 199, 215–16. Similar backgrounds are ascribed to other computer designers in Levy, *Hackers.*

16. Kidder, *Soul of a New Machine,* 30–32.

17. Ibid., 277.

18. Levy, *Hackers,* chaps. 8–13.

19. von Hippel, *Sources of Innovation,* chap. 6.

20. Ibid., 185. This is one of numerous short case studies found in the appendix to Von Hippel's book.

21. Gomory, "Ladder of Science," 101.

22. Mowrey and Rosenberg, *Pursuit of Economic Growth,* 230–35.

Bibliography

Books and Journals

Abernethy, John P. *The Modern Service of Commercial and Railway Telegraphy.* Cleveland: J. P. Abernethy, 1887.

Aitken, Hugh G. J. *The Continuous Wave: Technology and American Radio, 1900–1932.* Princeton: Princeton University Press, 1985.

———. *Scientific Management in Action: Taylorism at Watertown Arsenal, 1908–1915.* Princeton: Princeton University Press, 1985.

———. *Syntony and Spark: The Origins of Radio.* Princeton: Princeton University Press, 1976.

Aked, Charles K. "Alexander Bain: The Father of Electrical Horology." *Antiquarian Horology* 9 (1974): 52–60.

Akin, William. *Technocracy and the American Dream: The Technocratic Movement, 1900–1941.* Berkeley: University of California Press, 1977.

Andrews, Melodie. " 'What the Girls Can Do': The Debate Over the Employment of Women in the Early American Telegraph Industry." *Essays in Economic and Business History* 8 (1990): 109–20.

Appleby, Joyce. *Capitalism and a New Social Order: The Republican Vision of the 1790s.* New York: New York University Press, 1984.

Appleyard, Rollo. *The History of the Institution of Electrical Engineers (1871–1931).* London: Institution of Electrical Engineers, 1939.

Attali, Jacques, and Yves Stourdze. "The Birth of the Telephone and Economic Crisis: The Slow Death of the Monologue in French Society." In *The Social Impact of the Telephone,* ed. Ithiel de Sola Pool, 97–111. Cambridge: MIT Press, 1981.

Bailyn, Bernard. *The Ideological Origins of the American Revolution.* Cambridge: The Belknap Press of Harvard University Press, 1967.

Banning, Lance. "Jeffersonian Ideology Revisited: Liberal and Classical Ideas in the New American Republic." *William and Mary Quarterly* 43 (1986): 3–19.

Barclay, J. C. "Modern High Speed Printing Telegraph Systems." *Transactions of the International Electrical Congress, St. Louis, 1904* 3 (1904): 505–13.

Bell, John H. "Printing Telegraph Systems." *AIEE Transactions* 39 (1920): 167–230.

Bell Telephone Laboratories. *Engineering and Operations in the Bell System.* Murray Hill, N.J.: Bell Telephone Laboratories, 1977.

Bender, Thomas. *Community and Social Change in America.* New Brunswick, N.J.: Rutgers University Press, 1978.

————. *New York Intellect: A History of Intellectual Life in New York City, from 1750 to the Beginnings of Our Own Time.* New York: Knopf, 1987.

Beniger, James. *The Control Revolution: Technological and Economic Origins of the Information Society.* Cambridge: Harvard University Press, 1986.

Bidwell, Percy W. and John I. Falconer. *History of Agriculture in the Northern United States, 1620–1860.* New York: Peter Smith, 1941.

Birr, Kendall. "Industrial Research Laboratories." In *The Sciences in the American Context: New Perspectives,* ed. Nathan Reingold. Washington, D.C.: Smithsonian Institution Press, 1979.

Bliss, George. "The Telegraph Operator." *Journal of the Telegraph* 11 (1878): 242.

Blydenburgh, S. "On Facilitating the Advancement of Useful Improvements." *Mechanics' Magazine and Register of Inventions and Improvements* 2 (1833): 246.

Bode, H. W. *Synergy: Technical Integration and Technological Innovation in the Bell System.* Murray Hill, N.J.: Bell Telephone Laboratories, 1971.

Borut, Michael. "The *Scientific American* in Nineteenth-century America." Ph.D. diss., New York University, 1977.

Boston Directory. Boston: Adams, Sampson and Co., printed annually.

Braverman, Harry. *Labor and Monopoly Capital: The Degradation of Work in the Twentieth Century.* New York: Monthly Review Press, 1974.

Brock, Gerald W. *The Telecommunications Industry: The Dynamics of Market Structure.* Cambridge: Harvard University Press, 1981.

Brown, Richard D. *The Transformation of American Life, 1600–1865.* New York: Hill and Wang, 1976.

Bruce, Robert V. *Bell: Alexander Graham Bell and the Conquest of Solitude.* Boston: Little, Brown, 1973.

Butrica, Andrew J. "From *Inspecteur* to *Ingénieur*: Telegraphy and the Genesis of Electrical Engineering in France, 1845–1881." Ph.D. diss., Iowa State University, 1986.

————. "Telegraphy and the Genesis of Electrical Engineering Institutions in France, 1845–1895." *History and Technology* 3 (1987): 365–80.

Calahan, Edward A. "The District Telegraph." *Electrical World* (March 16, 1901). Typescript in Meadowcroft Box 4, ENHS.

————. "The Evolution of the Stock Ticker." *Electrical World* (February 9, 1901). Typescript in Meadowcroft Box 4, ENHS.

Calvert, Monte. *The Mechanical Engineer in America, 1830–1910: Professional Cultures in Conflict.* Baltimore: Johns Hopkins University Press, 1967.

Carlson, W. Bernard. "Invention, Science, and Business: The Professional Career of Elihu Thomson, 1870–1900." Ph.D. diss., University of Pennsylvania, 1984.

————. "The Myth of the Heroic Inventor and American Business." Paper presented at the Popular Culture Association Annual Meeting, Wichita, Kansas, April 1983.

Carosso, Vincent P. *Investment Banking in America: A History*. Cambridge: Harvard University Press, 1970.

Carroll, Charles F. "The Forest Society of New England." In *America's Wooden Age: Aspects of Its Early Technology,* ed. Brooke Hindle, 13–36. Tarrytown, N.Y.: Sleepy Hollow Restorations, 1975.

Cavanaugh, Cam, Barbara Hoskins, and Frances Pigeon. *At Speedwell Village in the Nineteenth Century*. Morristown, N.J.: Speedwell Village, 1981.

Cawelti, John. *Apostles of the Self-Made Man: Changing Concepts of Success in America*. Chicago: University of Chicago Press, 1965.

Chandler, Alfred P. *The Visible Hand: The Managerial Revolution in American Business*. Cambridge: The Belknap Press of Harvard University Press, 1977.

Chester, Stephen. *The American System of Fire-Alarm and Police Telegraph*. New York: Privately published, 1875. In Engineering Societies Library, New York.

Ciarlante, Marjorie H. "A Statistical Profile of Eminent American Inventors, 1700–1860: Social Origins and Roles." Ph.D. diss., Northwestern University, 1978.

Clemens, Paul. "The Operation of an Eighteenth-century Chesapeake Tobacco Plantation." *Agricultural History* 69 (1975): 517–31.

Cochran, Thomas, et al., eds. *The New American State Papers: Science and Technology*. Vol. 4, *Patents*. Wilmington, Del.: Scholarly Resources, 1973.

———. *The New American State Papers: Science and Technology*. Vol. 8, *Telegraphs, Military Technology*. Wilmington, Del.: Scholarly Resources, 1973.

Cooper, Carolyn. "The Role of Thomas Blanchard's Woodworking Inventions in Nineteenth-century American Manufacturing Technology." Ph.D. diss., Yale University, 1985.

Coulson, Thomas. *Joseph Henry: His Life and Work*. Princeton: Princeton University Press, 1950.

Craig, [Daniel H.]. *Craig's Manual of the Telegraph, Illustrating the Electro-Mechanical System of the American "Rapid" Telegraph Company of New York*. New York: John Polhemus, 1879.

———. *Machine Telegraph of Today*. New York: n.p., 1888.

Crehore, Albert, and George Squier. "The Synchronograph." *AIEE Transactions* 14 (1897): 93–154.

Czitrom, Daniel J. *Media and the American Mind from Morse to McLuhan*. Chapel Hill: University of North Carolina Press, 1982.

Danko, George. "Influence of Craft Method and Knowledge in Civil Engineering Practice." Paper delivered at the Annual Meeting of the Society for the History of Technology, Pittsburgh, October 1986.

Davies, Margery. *Women's Place Is at the Typewriter: Office Work and Office Workers, 1870–1930*. Philadelphia: Temple University Press, 1982.

Davis, Charles H., and Frank B. Rae. *Handbook of Electrical Diagrams and Connections.* New York: The Graphic Company, 1876.

Davis, Daniel, Jr. *Manual of Magnetism.* Boston: Daniel Davis, Jr., 1851.

Dawley, Alan. *Class and Community: The Industrial Revolution in Lynn.* Cambridge: Harvard University Press, 1976.

De Crèvecouer, J. Hector St. John. *Letters from an American Farmer and Sketches of Eighteenth-century America.* New York: Penguin Books, 1986.

Delany, Patrick. "Rapid Telegraphy." *Transactions of the International Electrical Congress, St. Louis, 1904* 3 (1904): 460–76.

———. "Telegraph Line Adjustment." *AIEE Transactions* 6 (1889): 506–13.

Dennis, Michael Aaron. "Accounting for Research: New Histories of Corporate Laboratories and the Social History of American Science." *Social Studies of Science* 17 (1987): 481–518.

Deyrup, Felicia J. *Arms Makers of the Connecticut Valley: A Regional Study of the Economic Development of the Small Arms Industry, 1798–1870.* Northampton, Mass: George Banta Publishing, 1948.

Dick, Thomas. "On the Improvement of Society by the Diffusion of Knowledge." *Mechanics' Magazine and Register of Inventions and Improvements* 2 (1833): 166–67.

Dictionary of American Biography. New York: Scribner's, 1964.

Dixon, James. "The Telautograph." *AIEE Transactions* 23 (1904): 645–57.

Dizard, Wilson P., Jr. *The Coming Information Age: An Overview of Technology, Economics, and Politics.* New York: Annenberg/Longman, 1982.

DuBoff, Richard B. "Business Demand and the Development of the Telegraph in the United States, 1844–1860." *Business History Review* 54 (1980): 459–79.

———. "The Telegraph and the Structure of Markets in the United States, 1845–1890." *Research in Economic History* 8 (1983): 253–77.

Dupree, A. Hunter. *Science in the Federal Government: A History of Policies and Activities to 1940.* Cambridge: Harvard University Press, 1957.

Fagen, M. D. *A History of Engineering and Science in the Bell System: The Early Years (1875–1925).* Murray Hill, N.J.: Bell Telephone Laboratories, 1975.

Faler, Paul G. *Mechanics and Manufacturers in the Early Industrial Revolution: Lynn, Massachusetts, 1780–1860.* Albany: State University of New York Press, 1981.

Ferguson, Eugene, ed. *Early Engineering Reminiscences (1815–40) of George Escol Sellers.* Washington, D.C.: Smithsonian Institution Press, 1965.

———. "The Mind's Eye: Nonverbal Thinking in Technology." *Science* 197 (1977): 827–36.

———. "On the Origin and Development of American Mechanical 'Know-How.'" *Midcontinent American Studies Journal* 3 (1962): 3–15.

Ferguson, Eugene, and Christopher Baer. *Little Machines: Patent Models in the Nineteenth Century.* Greenville, Del.: Hagley Museum, 1979.

Finn, William. "Wheatstone Automatic System." *Telegraph Age* 14 (1894). Typescript located in the Western Union Collection, National Museum of American History Archives, Smithsonian Institution.

Fischer, Henry C., and William H. Preece. "Joint Report upon the American Telegraph System" (1877). Manuscript in British Post Office Archives, London.

Fleming, A.P.M. *Industrial Research in the United States of America.* London: His Majesty's Stationery Office, 1917; rpt. New York: Arno Press, 1972.

Fletcher, Stephen W. *Pennsylvania Agriculture and Country Life, 1640–1840.* Harrisburg: Pennsylvania Historical and Museum Commission, 1950.

Flinn, John J. *History of the Chicago Police.* Chicago: Under the Auspices of the Police Fund, 1887.

Fowle, Frank. "Telegraph Transmission." *AIEE Transactions* 30 (1911): 1683–1741.

Friedel, Robert. *Pioneer Plastic: The Making and Selling of Celluloid.* Madison: University of Wisconsin Press, 1983.

Friedel, Robert, and Paul Israel. *Edison's Electric Light: Biography of an Invention.* New Brunswick, N.J.: Rutgers University Press, 1986.

Gabler, Edwin. *The American Telegrapher: A Social History, 1860–1900.* New Brunswick, N.J.: Rutgers University Press, 1988.

Gamewell Fire-Alarm Telegraph Company. *Fire-Alarm Telegraphs.* New York: Privately printed, 1907. Copy located at Division of Electricity, NMAH.

———. *Police Telephone and Signal Service.* Boston: Franklin Press: Rand, Avery, 1883. Copy located in Box 1048, AT&T.

Gartman, David. *Auto Slavery: The Labor Process in the American Automobile Industry, 1897–1950.* New Brunswick, N.J.: Rutgers University Press, 1986.

Gibb, George S. *The Saco-Lowell Shops: Textile Machinery Building in New England, 1813–1949.* Cambridge: Harvard University Press, 1950.

Gibson, F. M., and M.H.A. Lindsay. "Electric Protection Services: A Study of the Development of the Services Rendered by the American District Telegraph Company" (1962). Typescript in Division of Electricity, NMAH.

Gomory, Ralph E. "From the 'Ladder of Science' to the Product Development Cycle." *Harvard Business Review* 89 (1989): 99–105.

Gooding, David. " 'In Nature's School': Faraday as an Experimentalist." In *Faraday Rediscovered: Essays on the Life and Work of Michael Faraday, 1791–1867,* ed. David Gooding and Frank A.J.L. James, 105–35. New York: Stockton Press, 1985.

Goodrich, Carter. *Government Promotion of American Canals and Railroads, 1800–1890.* New York: Columbia University Press, 1960.

———. "The Revulsion against Internal Improvements." *Journal of Economic History* 10 (1950): 145–69.

Gordon, George N. *The Communications Revolution: A History of Mass Media in the United States.* New York: Hastings House, 1977.

Gray, Elisha. *Experimental Researches in Electro-Harmonic Telegraphy and Telephony.* New York: Russell Brothers, 1878.

Greene, John C. *American Science in the Age of Jefferson.* Ames: Iowa State University Press, 1984.

Greer, William. *A History of Alarm Security.* Washington, D.C.: National Burglar and Fire Alarm Association, 1979.

Gutman, Herbert. "The Reality of the Rags to Riches Myth: The Case of the Paterson, New Jersey, Locomotive, Iron, and Machinery Manufacturers, 1830–1880." In Gutman, *Work, Culture, and Society in Industrializing America* (New York: Vantage Books, 1977), 211–33.

Haber, Samuel. *Efficiency and Uplift: Scientific Management in the Progressive Era, 1890–1920.* Chicago: University of Chicago Press, 1964.

Hall, Peter Dobkin. *The Organization of American Culture, 1700–1900: Private Institutions, Elites, and the Origins of American Nationality.* New York: New York University Press, 1984.

Hall, Thomas. *Hall's Illustrated Catalogue of Telegraphic Instruments and Materials.* Boston: Privately printed, 1874.

Harlow, Alvin F. *Old Wires and New Waves: The History of the Telegraph, Telephone, and Wireless.* New York: Appleton-Century, 1936.

Harper, Douglas. *Working Knowledge: Skill and Community in a Small Shop.* Chicago: University of Chicago Press, 1987.

Hausmann, Erich. *Telegraph Engineering.* New York: Van Nostrand, 1915.

Hawley, Ellis W., ed. *Herbert Hoover as Secretary of Commerce: Studies in New Era Thought and Practice.* Iowa City: University of Iowa Press, 1981.

Hays, Samuel. *Conservation and the Gospel of Efficiency: The Progressive Conservation Movement, 1890–1920.* New York: Atheneum, 1979.

Healy, C. L. "Stock Tickers." *Telegraph Age* 23 (1905): 101–2, 120–22.

Higgs, Robert. "Urbanization and Inventiveness in the United States, 1870–1920." In *The New Urban History: Quantitative Explorations by American Historians,* ed. Leo F. Schnore, 247–59. Princeton: Princeton University Press, 1975.

Hindle, Brooke. *Emulation and Invention.* New York: New York University Press, 1981.

———. "From Art to Technology and Science." *Proceedings of the American Antiquarian Society* 96 (1986): 25–37.

———. *The Pursuit of Science in Revolutionary America, 1735–1789.* Chapel Hill: University of North Carolina Press, 1956.

Hindle, Brooke, and Steven Lubar. *Engines of Change: The American Industrial Revolution, 1790–1860.* Washington, D.C.: Smithsonian Institution Press, 1986.

Hirsch, Susan E. "From Artisan to Manufacturer: Industrialization and the

Small Producer in Newark, 1830–1860." In *Small Business in American Life,* ed. Stuart W. Bruchey, 80–99. New York: Columbia University Press, 1980.

———. *Roots of the American Working Class: The Industrialization of Crafts in Newark, 1800–1860.* Philadelphia: University of Pennsylvania Press, 1978.

Hoke, Donald R. *Ingenious Yankees: The Rise of the American System of Manufactures in the Private Sector.* New York: Columbia University Press, 1990.

Hotchkiss, Horace L. "The Stock Ticker." In *The New York Stock Exchange,* ed. Edmund Clarence Stedman, 433–41. New York: Greenwood Press, 1969.

Hounshell, David. "Edison and the Pure Science Ideal in Nineteenth-century America." *Science* 207 (1980): 612–17.

———. "Elisha Gray and the Telephone: On the Disadvantages of Being an Expert." *Technology and Culture* 16 (1975): 133–61.

———. *From the American System to Mass Production: The Development of Manufacturing Technology in the United States.* Baltimore: Johns Hopkins University Press, 1984.

———. "The Inventor as Hero in American History." Paper presented at Thomas Edison, Science, and Technology: A Commemorative Symposium, Newark, N.J., October 19–20, 1979.

———. "The Modernity of Menlo Park." In *Working at Inventing: Thomas A. Edison and the Menlo Park Experience,* ed. William S. Pretzer, 116–33. Dearborn, Mich.: Henry Ford Museum and Greenfield Village, 1989.

Houston, Edwin. "Delany's System of Facsimile Telegraph." *AIEE Transactions* 5 (1885): Paper no. 6.

———. "The Delany Synchronous Multi-Plex System of Telegraphy." *AIEE Transactions* (1884): Paper no. 5.

———. "Synchronism." *AIEE Transactions* 1 (1884): Paper no. 4.

Hughes, Thomas P. *American Genesis: A Century of Invention and Technological Enthusiasm, 1870–1970.* New York: Viking, 1989.

———. "Edison's Method." In *Technology at the Turning Point,* ed. William B. Pickett, 5–22. San Francisco: San Francisco Press, 1977.

———. *Elmer Sperry, Inventor and Engineer.* Baltimore: Johns Hopkins University Press, 1971.

———. "The Evolution of Large Technical Systems." In *The Social Construction of Technological Systems,* ed. Wiebe E. Bijker, Thomas P. Hughes, and Trevor Pinch, 51–82. Cambridge: MIT Press, 1989.

———. *Networks of Power: Electrification in Western Society, 1880–1930.* Baltimore: Johns Hopkins University Press, 1983.

———. "The Professional Inventor in the Heroic Age: Elmer Sperry in Brooklyn." In *Bridge to the Future: A Centennial Celebration of the Brooklyn Bridge,* ed. Margaret Latimer, Brooke Hindle, and Melvin Kranzberg, 163–70. New York: New York Academy of Sciences, 1984.

————. "Technological Momentum in History: Hydrogenation in Germany, 1898–1933." *Past and Present* 441 (1969): 106–32.

Hummel, Charles F. *With Hammer in Hand: The Dominy Craftsmen of East Hampton, New York.* Charlottesville: University Press of Virginia, 1968.

Hunt, Bruce. "'Practice vs. Theory': The British Electrical Debate, 1888–1891." *Isis* 74 (1983): 341–55.

Hunter, Louis C. *A History of Industrial Power in the United States, 1780–1930.* Vol. 1, *Water Power in the Century of the Steam Engine.* Charlottesville: University Press of Virginia, 1979.

————. *A History of Industrial Power in the United States, 1780–1930.* Vol. 2, *Steam Power.* Charlottesville: University Press of Virginia, 1985.

Huntington, F. M. *The Telegrapher's Souvenir. A Work Comprising Compilations and Original Articles . . . Intended to Be Instructive, Interesting and Amusing. Not Only to the Fraternity, But to Strangers as Well.* Paterson, N.J.: Lyon and Halsted, 1875.

Hurst, James Willard. *Law and the Conditions of Freedom in the Nineteenth-century United States.* Madison: University of Wisconsin Press, 1956.

————. *The Legitimacy of the Business Corporation in the Law of the United States, 1780–1970.* Charlottesville: University Press of Virginia, 1970.

Hyde, Charles K. "Iron and Steel Technologies Moving between Europe and the United States, before 1914." In *International Technology Transfer: Europe, Japan, and the USA, 1700–1914,* ed. David J. Jeremy, 51–73. Aldershot, Eng.: Edward Elgar Publishing, 1991.

Innes, Stephen. *Labor in a New Land: Economy and Society in Seventeenth-century Springfield.* Princeton: Princeton University Press, 1983.

Israel, Paul B. "Telegraphy and Edison's Invention Factory." In *Working at Inventing: Thomas A. Edison and the Menlo Park Experience,* ed. William S. Pretzer, 66–83. Dearborn, Mich.: Henry Ford Museum and Greenfield Village, 1989.

Israel, Paul B., and Keith A. Nier. "The Transfer of Telegraph Technologies in the Nineteenth Century." In *International Technology Transfer: Europe, Japan, and the USA, 1700–1914,* ed. David J. Jeremy, 95–121. Aldershot, Eng.: Edward Elgar Publishing, 1991.

Jacob, Margaret. *The Cultural Meaning of the Scientific Revolution.* New York: Knopf, 1988.

Jenkins, Reese. *Images and Enterprise: Technology and the American Photographic Industry, 1839–1925.* Baltimore: Johns Hopkins University Press, 1975.

Jeremy, David J. *Transatlantic Industrial Revolution: The Diffusion of Textile Technologies between Britain and America, 1790–1830s.* Cambridge: MIT Press, 1981.

Jeremy, David J., and Darwin H. Stapleton. "Transfers between Culturally Related Nations: The Movement of Textile and Railroad Technologies between Britain and the United States, 1780–1840." In *International Tech-*

nology Transfer: Europe, Japan, and the USA, 1700–1914, ed. Jeremy, 31–48. Aldershot, Eng.: Edward Elgar Publishing, 1991.

John, Richard R., Jr. "A Failure of Vision? The Jacksonians, the Post Office, and the Telegraph, 1844–1847." Paper presented at the Society for the History of Technology Annual Meeting, Pittsburgh, October 1986.

———. "The Origins of Commercial Telegraphy in the United States, 1844–1847." Paper presented at the Economic History Workshop, September 1987.

Johnston, William John. *Telegraphic Tales and Telegraphic History.* New York: William J. Johnston, 1877.

Jones, Francis W. "The Quadruplex." *Journal of the American Electrical Society* (1875): 16–29.

———. "Static Difficulties in Telegraph Wires." *AIEE Transactions* 3 (1886): 120–27.

Josephson, Matthew. *Edison: A Biography.* New York: McGraw-Hill, 1959.

Jungnickel, Christina, and Russell McCormmach. *The Intellectual Mastery of Nature: Theoretical Physics from Ohm to Einstein.* Vol. 1, *The Torch of Mathematics.* Chicago: University of Chicago Press, 1986.

Kaempffert, Waldemar. *Invention and Society.* Chicago: American Library Association, 1930.

———. "Invention as a Social Manifestation." In *A Century of Progress,* ed. Charles A. Beard, 21–65. New York: Harper and Brothers, 1932.

———. *A Popular History of American Invention.* New York: Scribner's, 1924.

Kargon, Robert H. *Science in Victorian Manchester: Enterprise and Expertise.* Baltimore: Johns Hopkins University Press, 1977.

Kasson, John F. *Civilizing the Machine: Technology and Republican Values in America, 1776–1900.* New York: Penguin Books, 1977.

Kerber, Linda K. "The Republican Ideology of the Revolutionary Generation." *American Quarterly* 37 (1985): 474–95.

Kevles, Daniel J. *The Physicists: The History of a Scientific Community in Modern America.* New York: Vantage Books, 1979.

Kidder, Tracy. *The Soul of a New Machine.* New York: Avon Books, 1981.

Kieve, Jeffrey. *Electric Telegraph in the U.K.: A Social and Economic History.* New York: Barnes and Noble, 1973.

Kinsley, Carl. "A High-Speed Printing Telegraph System." *AIEE Transactions* 33 (1914): 1243–53.

Klein, Maury. *The Life and Legend of Jay Gould.* Baltimore: Johns Hopkins University Press, 1986.

Kline, Ronald. "Engineering R&D at the Westinghouse Electric Company, 1886–1922." Paper presented at the Society for the History of Technology Annual Meeting, Pittsburgh, October 1986.

———. "Science and Engineering Theory in the Invention and Development of the Induction Motor, 1880–1900." *Technology and Culture* 28 (1987): 283–313.

Kloppenberg, James T. "The Virtues of Liberalism: Christianity, Republicanism, and Ethics in Early American Political Discourse." *Journal of American History* 74 (1987): 9–33.

Kornblith, Gary K. "The Craftsman as Industrialist: Jonas Chickering and the Transformation of American Piano Making." *Business History Review* 59 (1985): 349–68.

Kranzberg, Melvin. "Broadening and Deepening U.S. Engineering Curriculum." In *Technological Education—Technological Style,* ed. Kranzberg, 75–84. San Francisco: San Francisco Press, 1986.

Landes, David S. "Hand and Mind in Time Measurement: The Contributions of Art and Science." In *Innovation at the Crossroads between Science and Technology,* ed. Melvin Kranzberg, et al., 81–93. Haifa, Israel: S. Neaman Press, 1989.

Langdale, John. "Impact of the Telegraph on the Buffalo Agricultural Commodity Market: 1846–1848." *Professional Geographer* 31 (1979): 165–69.

Laurie, Bruce. *Working People of Philadelphia, 1800–1850.* Philadelphia: Temple University Press, 1980.

Laurie, Bruce, and Mark Schmitz. "Manufacture and Productivity: The Making of an Industrial Base, Philadelphia, 1850–1880." In *Philadelphia: Work, Space, Family, and Group Experience in the Nineteenth Century,* ed. Theodore Hershberg, 43–92. New York: Oxford University Press, 1981.

Laurie, Bruce, George Alter, and Theodore Hershberg. "Immigrants and Industry: The Philadelphia Experience, 1850–1880." In *Philadelphia: Work, Space, Family, and Group Experience in the Nineteenth Century,* ed. Theodore Hershberg, 93–119. New York: Oxford University Press, 1981.

Layton, Edwin. "American Ideologies of Science and Engineering." *Technology and Culture* 17 (1976): 688–701.

———. "Mirror-Image Twins: The Communities of Science and Technology in Nineteenth-century America." *Technology and Culture* 12 (1971): 562–80.

———. *The Revolt of the Engineers: Social Responsibility and the American Engineering Profession.* Baltimore: Johns Hopkins University Press, 1986.

Leahey, Philip. "Skilled Labor and the Rise of the Modern Corporation: The Case of the Electrical Industry." *Labor History* 27 (1985–86): 31–53.

Leary, Thomas E. "Industrial Ecology and the Labor Process: The Redefinition of Craft in New England Textile Machinery Shops." In *Life and Labor: Dimensions of American Working-Class History,* ed. Charles Stephenson and Robert Asher, 37–56. Albany: State University of New York Press, 1986.

Lefferts, Marshall. "The Electric Telegraph: Its Influence and Geographical Distribution." American Geographical and Statistical Society *Bulletin* 2 (1857): 242–64.

Lemon, James T. *The Best Poor Man's Country: A Geographical Study of Early Southeastern Pennsylvania.* New York: W. W. Norton, 1972.

Leslie, Stuart W. *Boss Kettering: Wizard of General Motors.* New York: Columbia University Press, 1983.

Letwin, William. *Law and Economic Policy in America: The Evolution of the Sherman Antitrust Act.* New York: Random House, 1965.

Levy, Steven. *Hackers: Heroes of the Computer Revolution.* Garden City, N.Y.: Doubleday, 1984.

Lewis, W. David. "Industrial Research and Development." In *Technology in Western Civilization,* vol. 2, ed. Melvin Kranzberg and Carroll W. Pursell, Jr., 615–34. New York: Oxford University Press, 1967.

Licht, Walter. "The Dialectics of Bureaucratization: The Case of Nineteenth-century American Railway Workers." In *Life and Labor: Dimensions of American Working-Class History,* ed. Charles Stephenson and Robert Asher, 92–114. Albany: State University of New York Press, 1986.

———. *Working for the Railroad: The Organization of Work in the Nineteenth Century.* Princeton: Princeton University Press, 1983.

Lindley, Lester G. *The Constitution Faces Technology: The Relationship of the National Government to the Telegraph, 1866–1884.* New York: Arno Press, 1975.

Lockwood, Thomas D. *Electricity, Magnetism, and Electric Telegraphy: A Practical Guide and Hand-Book.* 2d ed. New York: D. Van Nostrand, 1888.

Lozier, John W. *Taunton and Mason: Cotton Machinery and Locomotive Manufacture in Taunton, Massachusetts, 1811–1861.* New York: Garland Publishing, 1986.

Lubar, Steven. "Corporate and Urban Contexts of Textile Technology in Nineteenth-century Lowell, Massachusetts: A Study of the Social Nature of Technological Knowledge." Ph.D. diss., University of Chicago, 1983.

———. "Cultural and Technological Design in the Nineteenth-century Pin Industry: John Howe and the Howe Manufacturing Company." *Technology and Culture* 28 (1987): 253–82.

Mabee, Carleton. *The American Leonardo: A Life of Samuel F. B. Morse.* New York: Knopf, 1943.

McCoy, Drew R. *The Elusive Republic: Political Economy in Jeffersonian America.* New York: W. W. Norton, 1980.

McCusker, John J., and Russell R. Menard. *The Economy of British America, 1607–1789.* Chapel Hill: University of North Carolina Press, 1985.

McGaw, Judith A. *Most Wonderful Machine: Mechanization and Social Change in Berkshire Paper Making, 1801–1885.* Princeton: Princeton University Press, 1987.

McMahon, Michal. "Corporate Technology: The Origins of the American Institute of Electrical Engineers." *IEEE Proceedings* 64 (1976): 1383–90.

————. *The Making of a Profession: A Century of Electrical Engineering in America.* New York: IEEE Press, 1984.

McNicol, Donald. *American Telegraph Practice.* New York: McGraw-Hill, 1913.

Martin, James. *The Wired Society.* Englewood Cliffs, N.J.: Prentice-Hall, 1978.

Marx, Leo. *The Machine in the Garden: Technology and the Pastoral Ideal in America.* New York: Oxford University Press, 1964.

Matson, Cathy, and Peter Onuf. "Towards a Republican Empire: Interest and Ideology in Revolutionary America." *American Quarterly* 37 (1985): 497–531.

Mau, Clayton C. "The Early History of the Telegraph in the United States." Ph.D. diss., Cornell University, 1930.

Maver, William. *American Telegraphy: Systems, Apparatus, Operation.* New York: J. H. Bunnell, 1892.

————. *American Telegraphy and Encyclopedia of the Telegraph: Systems, Apparatus, Operation,* rev. ed. New York: Maver Publishing, 1903.

Maver, William, and Donald McNicol. "American Telegraph Engineering— Notes on History and Practice." *AIEE Transactions* 29 (1910): 1303–56.

Mayntz, Renate, and Thomas P. Hughes, eds. *The Development of Large Technical Systems.* Boulder, Colo.: Westview Press, 1988.

Medbery, James K. *Men and Mysteries of Wall Street.* 1879; Rpt.: New York: Greenwood Press, 1968.

Meier, Hugo A. "Technology and Democracy, 1800–1860." *Mississippi Valley Historical Review* 43 (1957): 618–40.

————. "The Technological Concept in American Social History, 1750–1860." Ph.D. diss., University of Wisconsin–Madison, 1950.

————. "Thomas Jefferson and a Democratic Technology." In *Technology in America: A History of Individuals and Values,* ed. Carroll W. Pursell, Jr., 17–44. Cambridge: MIT Press, 1982.

Meier, Richard L. "The Organization of Technical Innovation in Urban Environments." In *The Historian and the City,* ed. Oscar Handlin and John Burchard, 74–83. Cambridge: MIT Press and Harvard University Press, 1963.

Middlekauf, Robert. *The Mathers: Three Generations of Puritan Intellectuals, 1596–1728.* New York: Oxford University Press, 1971.

Millard, Andre. *Edison and the Business of Innovation.* Baltimore: Johns Hopkins University Press, 1990.

Mitchell, Robert D. *Commercialism and Frontier: Perspectives on the Early Shenandoah Valley.* Charlottesville: University Press of Virginia, 1977.

Mollela, Arthur, and Nathan Reingold. "Theorists and Ingenious Mechanics: Joseph Henry Defines Science." *Science Studies* 3 (1973): 323–51.

Montgomery, David. *Fall of the House of Labor.* Cambridge: Cambridge University Press, 1987.

Moreau, Louise Ramsey. "The Story of the Key." *Morsum Magnificat* 6 (Winter 1987).

Morison, Elting. *From Know-How to Nowhere: The Development of American Technology.* New York: Basic Books, 1974.

Morse, Samuel F. B. *Letters and Journals.* Edited by Edward Lind Morse. New York: Da Capo Press, 1973.

Mowery, David C., and Nathan Rosenberg. *Technology and the Pursuit of Economic Growth.* Cambridge: Cambridge University Press, 1989.

Musson, A. E., and Eric Robinson. *Science and Technology in the Industrial Revolution.* Toronto: University of Toronto Press, 1969.

National Research Council. *Research Laboratories in Industrial Establishments of the United States.* Washington, D.C.: Government Printing Office, 1920, 1921, 1927, 1931.

Nelson, Daniel. *Frederick W. Taylor and the Rise of Scientific Management.* Madison: University of Wisconsin Press, 1980.

————. *Managers and Workers: Origins of the New Factory System in the United States, 1880–1920.* Madison: University of Wisconsin Press, 1975.

Newhall, Beaumont. *The Daguerreotype in America.* New York: New York Graphic Society, 1961.

Noble, David. *America by Design: Science, Technology, and the Rise of Corporate Capitalism.* New York: Knopf, 1977.

Parsons, Frank. *The Telegraph Monopoly.* Philadelphia: C. F. Taylor, 1889.

Phillips, Walter Polk. *Oakum Pickings: A Collection of Stories, Sketches, and Paragraphs Contributed from Time to Time to the Telegraphic and General Press.* New York: W. J. Johnston, 1876.

Phillips, Willard. *The Law of Patents for Inventions.* Boston: S. Colman and Collins, Keese, 1837.

Pond, Chester. *An Outline of Theoretical Telegraphy.* Oberlin, Ohio: Calkins, Griffin, 1866.

Pope, Franklin L. "The American Inventors of the Telegraph, with Special Attention to the Services of Alfred Vail." *Century Magazine* 35 (1888): 924–44.

————. "The Invention of the Electromagnetic Telegraph—II." *Electrical World* 26 (1895): 181–85.

————. *Modern Practice of the Electric Telegraph.* New York: Russell Brothers, 1869.

————. "The Western Union Telegraph Company's Manufactory." *Telegrapher* 8 (1871): 104–5.

Pope, Ralph W. "The Rise of the Stock Reporting Telegraph." *Electrical World* (March 4, 1899). Typescript in Meadowcroft Box 4, ENHS.

Post, Robert C. "'Liberalizers' versus 'Scientific Men' in the Antebellum Patent Office." *Technology and Culture* 17 (1976): 24–54.

————. *Physics, Patents, and Politics: A Biography of Charles Grafton Page.* New York: Science History Publications, 1976.

Potts, Louis M. "Printing Telegraphy." *Transactions of the International Electrical Congress, St. Louis, 1904* 3 (1904): 477–92.

————. "The Rowland Telegraphic System." *AIEE Transactions* 26 (1907): 539–46.

Pred, Allan. *The Spatial Dynamics of U.S. Urban-Industrial Growth, 1800–1914: Interpretive and Theoretical Essays.* Cambridge: MIT Press, 1966.

————. *Urban Growth and City Systems in the United States, 1840–1860.* Cambridge: Harvard University Press, 1980.

Preece, William. "On the Advantages of Scientific Education: A Lecture Addressed to the Telegraph Staff." *Journal of the Society of Telegraph Engineers* 1 (1871): 266–76.

————. *Recent Advances in Telegraphy.* London: Society for the Encouragement of Arts, Manufactures, and Commerce, 1879.

Prescott, George. *Electricity and the Electric Telegraph.* New York: D. Appleton, 1877; rev. ed., 1885.

————. *History, Theory, and Practice of the Electric Telegraph.* Boston: Ticknor and Fields, 1863; rev. ed., 1866.

Preston, Daniel. "The Administration and Reform of the U.S. Patent Office, 1790–1836." *Journal of the Early Republic* 5 (1985): 331–48.

Prime, Samuel I. *The Life of Samuel F. B. Morse.* New York: Arno Press, 1974.

Prindle, Edwin J. *Patents as a Factor in Manufacturing.* New York: Engineering Magazine, 1908.

Pursell, Carroll, Jr. *Early Stationary Steam Engines in America: A Study in the Migration of a Technology.* Washington, D.C.: Smithsonian Institution Press, 1969.

————. "Thomas Digges and William Pearce: An Example of the Transit of Technology." *William and Mary Quarterly* 21 (1964): 551–60.

————. "Women Inventors in America." *Technology and Culture* 22 (1981): 545–49.

Rae, John. "The Application of Science to Industry." In *The Organization of Knowledge in Modern America, 1860–1920,* ed. Alexandra Oleson and John Voss, 249–68. Baltimore: Johns Hopkins University Press, 1979.

Rawson, Marion N. *Little Old Mills.* New York: E. P. Dutton, 1935.

Reader, W. J. *A History of the Institution of Electrical Engineers, 1871–1971.* London: Peter Pregrinus, 1987.

Reiber, A. H. "Printing Telegraph Systems Applied to Message Traffic Handling." *AIEE Transactions* 41 (1922): 39–51.

Reich, Leonard S. "Irving Langmuir and the Pursuit of Science and Technology in the Corporate Environment." *Technology and Culture* 24 (1983): 199–221.

————. *The Making of American Industrial Research: Science and Business at GE and Bell, 1876–1926.* Cambridge: Cambridge University Press, 1985.

————. "Research, Patents, and the Struggle to Control Radio: A Study of Big Business and the Uses of Industrial Research." *Business History Review* 51 (1977): 233–34.

Reid, James. *The Telegraph in America.* New York: Derby Brothers, 1879; rev. ed.: New York: John Polhemus, 1886.

Reingold, Nathan, et al., eds. *The Papers of Joseph Henry:* Vol. 2, *The Princeton Years.* Washington, D.C.: Smithsonian Institution Press, 1975.

Rhoads, Stanley. "Some Phases of Railroad Telegraph and Telephone Engineering." *AIEE Transactions* 30 (1921): 301–80.

Robinson, Eric. "The Derby Philosophical Society." *Annals of Science* 9 (1953): 359–67.

Rock, Howard R. *Artisans of the New Republic: The Tradesmen of New York City in the Age of Jefferson.* New York: New York University Press, 1979.

Rorbaugh, W. J. *The Craft Apprentice: From Franklin to the Machine Age in America.* New York: Oxford University Press, 1986.

Rosenberg, Nathan. "America's Rise to Woodworking Leadership." In *America's Wooden Age: An Aspect of Its Early Technology,* ed. Brooke Hindle, 37–62. Tarrytown, N.Y.: Sleepy Hollow Restorations, 1975.

———. *Perspectives on Technology.* Cambridge: Cambridge University Press, 1976.

———. "Technological Change in the Machine Tool Industry, 1840–1910." *Journal of Economic History* 23 (1963): 414–43.

Rosenberg, Robert. "Academic Physics and the Origins of Electrical Engineering in America." Ph.D. diss., Johns Hopkins University, 1990.

———. "American Physics and the Origins of Electrical Engineering." *Physics Today* 36 (1983): 48–54.

———. "Test Men, Experts, Brother Engineers, and Members of the Fraternity: Whence the Early Electrical Work Force?" *IEEE Transactions on Education* 27 (1984): 203–10.

Rosenberg, Robert, and Paul Israel. "Intraurban Telegraphy: The Nerve of Some Cities." Paper presented at American Historical Association Annual Meeting, Chicago, December 1986.

St. George, Robert Blair. "Fathers, Sons, and Identity: Woodworking Artisans in Southeastern New England, 1620–1700." In *The Craftsman in Early America,* ed. Ian M. G. Quimby, 89–125. New York: W. W. Norton, 1984.

Samson [pseud.]. *Sam Johnson; The Experience and Observations of a Railroad Telegraph Operator.* New York: W. J. Johnston, 1878.

Scharlott, Bradford W. "The Telegraph and the Integration of the U.S. Economy: The Impact of Electrical Communications on Interregional Prices and the Commercial Life of Cincinnati." Ph.D. diss., University of Wisconsin–Madison, 1986.

Schatz, Ronald W. *The Electrical Workers: A History of Labor at General Electric and Westinghouse, 1923–60.* Urbana: University of Illinois Press, 1983.

Schofield, Robert. *The Lunar Society of Birmingham.* Oxford: Clarendon Press, 1963.

Scranton, Philip. "Learning Manufacture: Education and Shop-Floor Schooling in the Family Firm." *Technology and Culture* 27 (1986): 40–62.

Scribner, Charles E. "Log-Fire Reminiscences of a Pioneer." *Western Electric News* 13 (1925): 2–5, 21–24, 27–30, 39–42.

Seavoy, Ronald E. *The Origins of the American Business Corporation, 1784–1855: Broadening the Concept of Public Service during Industrialization.* Westport, Conn.: Greenwood Press, 1982.

Seidel, Rita, "Von der electrischen Telegraphie zur Elektrotechnik. Zur Genese einer wissenschaftlichen Disziplin. Hanover als Beispiel." *Zeitschrift der Universität Hannover* 11 (1984): 39–47.

Shaffner, Taliaferro P. *The Telegraph Manual.* New York: Pudney and Russell, 1859.

Shallope, Robert E. "Toward a Republican Synthesis: The Emergence of an Understanding of Republicanism in American Historiography." *William and Mary Quarterly* 29 (1972): 49–80.

Sinclair, Bruce. *A Centennial History of the American Society of Mechanical Engineers, 1880–1980.* Toronto: University of Toronto Press, 1980.

———. *Philadelphia's Philosopher Mechanics: A History of the Franklin Institute, 1824–65.* Baltimore: Johns Hopkins University Press, 1974.

Sklar, Martin J. *The Corporate Reconstruction of American Capitalism, 1890–1916.* Cambridge: Cambridge University Press, 1988.

Sloane, Eric. *Diary of an Early American Boy.* New York: Wilfred Funk, 1962.

Smith, George David. *The Anatomy of a Business Strategy: Bell, Western Electric, and the Origins of the American Telephone Industry.* Baltimore: Johns Hopkins University Press, 1985.

Smith, John Kenly, Jr. "The Scientific Tradition in American Industrial Research." *Technology and Culture* 31 (1990): 121–31.

Smith, Merrit Roe. "John H. Hall, Simeon North, and the Milling Machine: The Nature of Innovation among Antebellum Arms Makers." *Technology and Culture* 14 (1973): 573–91.

Stanley, Autumn. "The Patent Office Clerk as Conjurer." In *Women, Work, and Technology: Transformations,* ed. Barbara D. Wright, et al., 118–36. Ann Arbor: University of Michigan Press, 1987.

Stapleton, Darwin H. "Early Industrial Research in Cleveland: Local Networks and National Connections." Paper presented at the Organization of American Historians Annual Meeting, Philadelphia, April 1987.

———. *The Transfer of Industrial Technologies to Early America.* Philadelphia: American Philosophical Society, 1986.

Steffens, Charles G. *The Mechanics of Baltimore: Workers and Politics in the Age of Revolution, 1763–1812.* Urbana: University of Illinois Press, 1984.

Taltavall, John B. *Telegraphers of To-day: Descriptive, Historical, Biographical.* New York: John B. Taltavall, 1893.

Tarr, Joel A., with Thomas Finholt and David Goodman. "The City and the

Telegraph: Urban Telecommunications in the Pre-Telephone Era." *Journal of Urban History* 14 (1987): 38–80.

Taylor, William B. "An Historical Sketch of Henry's Contribution to the Electro-Magnetic Telegraph." *Smithsonian Institution Report for 1878.* Rpt. in *The Electric Telegraph: An Historical Anthology,* ed. George Shiers. New York: Arno Press, 1977.

Teaford, Jon C. *The Unheralded Triumph: City Government in America, 1870–1900.* Baltimore: Johns Hopkins University Press, 1984.

Thompson, William. *Wiring a Continent: The History of the Telegraph Industry in the United States, 1832–1866.* Princeton: Princeton University Press, 1947.

Thomson, Ross. *The Path to Mechanized Shoe Production in the United States.* Chapel Hill: University of North Carolina Press, 1989.

Thornton, William. "Account of the Method of Obtaining Patents." *Emporium of Arts and Sciences* 2 (1813): 228–31.

Timmons, George. "Education and Technology in the Industrial Revolution." *History of Technology* 8 (1983): 135–49.

Toner, J. M. "Washington as an Inventor and Promoter of the Useful Arts." In *Patent Centennial Celebration, 1891: Proceedings and Addresses,* 313–79. Washington, D.C.: Press of Gedney and Roberts, 1892.

Tosiello, Rosario J. *The Birth and Early Years of the Bell Telephone System, 1876–1880.* New York: Arno Press, 1979.

Trachtenberg, Alan. *The Incorporation of America: Culture and Society in the Gilded Age.* New York: Hill and Wang, 1982.

Turnbull, L. *The Electro-Magnetic Telegraph.* Philadelphia: A. Hart, 1852.

Ulriksson, Vidkunn. *The Telegraphers: Their Craft and Union.* Washington, D.C.: Public Affairs Press, 1953.

U.S. Bureau of the Census. *Census Reports Volume X, Manufactures Part IV.* Washington, D.C.: Government Printing Office, 1902.

———. *Historical Statistics of the United States, Colonial Times to the Present.* Washington, D.C.: Government Printing Office, 1975.

———. *Special Reports of the Census Office, Manufacturers Part IV, Selected Industries.* Washington, D.C.: Government Printing Office, 1905.

U.S. Industrial Commission. *Report on Transportation, Volume IX of the Commission's Reports.* Washington, D.C.: Government Printing Office, 1901.

U.S. Postmaster General. *An Argument in Support of the Limited Post and Telegraph.* Washington, D.C.: Government Printing Office, 1890.

Usselman, Steven W. "Running the Machine: The Management of Technological Innovation on American Railroads, 1860–1910." Ph.D. diss., University of Delaware, 1984.

Vail, Alfred. *The Electro Magnetic Telegraph.* Philadelphia: Lea and Blanchard, 1845.

Vail, J. Cummings. *Early History of the Electro-Magnetic Telegraph, from Letters and Journals of Alfred Vail.* New York: Hine Brothers, 1914.

Vansize, William B. "A New Page-Printing Telegraph." *AIEE Transactions* 18 (1901): 37–43.

Varley, Cromwell Fleetwood. "Report on the Condition of the Lines of the Western Union Telegraph Company." New York: Western Union, 1867. In Engineering Societies Library, New York.

Verplanck, Gulian C. "Introductory Address Delivered before the Mechanics' Institute of the City of New-York." *Mechanics' Magazine and Register of Inventions and Improvements* 3 (1834): 47–49.

von Fischer-Treuenfeld, R. "Fire Telegraphs." *Journal of the Society of Telegraph Engineers* 6 (1877): 75–121.

———. "The Present State of Fire-Telegraphy." *Journal of the Society of Telegraph Engineers* 17 (1888): 258–307.

von Hippel, Erich. *The Sources of Innovation*. New York: Oxford University Press, 1988.

Wallace, Anthony C. *Rockdale: The Growth of an American Village in the Early American Industrial Revolution*. New York: W. W. Norton, 1972.

Wasserman, Neil H. *From Invention to Innovation: Long-Distance Telephone Transmission at the Turn of the Century*. Baltimore: Johns Hopkins University Press, 1985.

Watson, Thomas. *Exploring Life: The Autobiography of Thomas Watson*. New York: D. Appleton, 1926.

Watts, Steven. *The Republic Reborn: War and the Making of Liberal America, 1790–1820*. Baltimore: Johns Hopkins University Press, 1987.

Weinstein, James. *The Corporate Ideal in the Liberal State, 1900–1918*. Boston: Beacon Press, 1968.

Weiss, Harry B., and Grace M. Weiss. *The Early Sawmills of New Jersey*. Trenton: New Jersey Agricultural Society, 1968.

———. *Forgotten Mills of Early New Jersey*. Trenton: New Jersey Agricultural Society, 1960.

Welch, Walter. *Charles Batchelor: Edison's Chief Partner*. Syracuse, N.Y.: Syracuse University, 1972.

Welter, Rush. *The Mind of America, 1820–1869*. New York: Columbia University Press, 1975.

Western Union Telegraph Company. *Annual Reports* (printed annually). WUC.

———. Engineering Department. "A Brief Outline of the Technical Progress Made by the Western Union Telegraph" (1935). WUC.

———. *The Proposed Union of the Telegraph and Postal Systems*. Cambridge, Mass.: Welch, Bigelow, 1869.

———. *A Review of the "Opinion of Experts" as to the Necessity for the Poles now erected in Tenth Street, in the City of Philadelphia by the Western Union Telegraph Company*. New York: Russell Brothers, 1876. In Western Union Engineering Library.

———. *Statement of the Western Union Telegraph Company on the Proposed*

Union of the Telegraph and Postal Systems. Cambridge, Mass.: Welch, Bigelow, 1869.

Whalley, Peter. "The Social Practice of Independent Inventing." *Science, Technology, and Human Values* 16 (1991): 208–32.

White, John H. *A History of the American Locomotive, Its Development: 1830–1880*. New York: Dover Publications, 1968.

Whitney, James A. *The Relationship of the Patent Laws to American Agriculture, Arts, and Industries*. New York: New York Society of Practical Engineering, 1875.

Wiebe, Robert. *The Search for Order, 1877–1920*. New York: Hill and Wang, 1967.

Wilentz, Sean. *Chants Democratic: New York City and the Rise of the American Working Class, 1788–1850*. New York: Oxford University Press, 1984.

Williams, Frederick. *The Communications Revolution*. New York: New American Library, 1983.

———. *Technology and Communication Behavior*. Belmont, Calif.: Wadsworth Publishing, 1987.

Wise, George. "A New Role for Professional Scientists in Industry: Industrial Research at General Electric, 1900–1916." *Technology and Culture* 21 (1980): 408–29.

———. *Willis R. Whitney, General Electric, and the Origins of U.S. Industrial Research*. New York: Columbia University Press, 1985.

Wish, Judith Barry. "From Yeoman Farmer to Industrious Producer: The Relationship between Classical Republicanism and the Development of Manufacturing in America from the Revolution to 1850." Ph.D. diss., Washington University, 1976.

Wood, Gordon S. *The Creation of the American Republic, 1776–1787*. Chapel Hill: University of North Carolina Press, 1969.

Wylie, Irvin. *The Self-Made Man in America: The Myth of Rags to Riches*. New York: Free Press, 1954.

York, Neil. *Mechanical Metamorphosis: Technological Change in Revolutionary America*. Westport, Conn.: Greenwood Press, 1985.

Patent Interference Files

Anders v. Warner. Patent Interference File 5603. U.S. Patent Office Records, MdSuFR.

Buckingham v. Jones. Patent Interference File 10,588. U.S. Patent Office Records, MdSuFR.

Buell v. Martin. Patent Interference File 10,110. U.S. Patent Office Records, MdSuFR.

Cochrane v. Van Hoevenbergh. Patent Interference File 7954. U.S. Patent Office Records, MdSuFR.

Edison v. Lane v. Gray v. Rose v. Gilliland. Patent Interference Files 8027 and 8028. U.S. Patent Office Records, MdSuFR.

Field v. Lugo. Patent Interference File 8087. U.S. Patent Office Records, MdSuFR.

Field v. Pope. Patent Interference File 7976. U.S. Patent Office Records, MdSuFR.

Gamewell v. Chester. Patent Interference File 5406. U.S. Patent Office Records, MdSuFR.

Gamewell & Co. v. Allen. Patent Interference File 360. U.S. Patent Office Records, MdSuFR.

Ladd v. Seiler. Patent Interference File 8765. U.S. Patent Office Records, MdSuFR.

McCullough v. Watkins. Patent Interference File 5081. U.S. Patent Office Records, MdSuFR.

Nicholson v. Edison. Patent Interference Files 8689 and 8690. U.S. Patent Office Records, MdSuFR.

Phelps v. Anders. Patent Interference Files 3993 and 6739. U.S. Patent Office Records, MdSuFR.

Ruddick v. Milliken. Patent Interference File 15,788. MdSuFR.

"Telephone Interferences." A combined group of interference proceedings that became known as the Telephone Interferences. The entire series of interferences is at MdSuFR; parts are also found at ENHS.

Von Hefner Alteneck v. Western Union. ENHS.

Wiley v. Field. Patent Interference File 9219. U.S. Patent Office Records, MdSuFR.

Wilson v. MacKenzie. Patent Interference File 9310. U.S. Patent Office Records, MdSuFR.

United States District Court Cases

Boston Safe Deposit Co. et al. v. American Rapid Telegraph Co. et al. Equity Case File 3389. Southern District of New York. NjBaFAR.

Boston Safe Deposit Co. et al. v. Bankers and Merchants Telegraph Co. et al. Equity Case File 4314. Southern District of New York. NjBaFAR.

Commercial Telegraph Co. v. Gold and Stock Telegraph Co. Equity Case File 2893. Southern District of New York. NjBaFAR.

Edison Electric Light Co. v. F. P. Little Electrical Construction and Supply Co. et al. Equity Case File 6204. Northern District of New York. ENHS.

French v. Rogers. Equity Case File 104. U.S. Circuit Court, Eastern District of Pennsylvania. Philadelphia: U.S. Steam Power and Job Printing Office, 1851.

Gamewell et al. v. Chester and Chester. Equity Case File 183. Southern District of New York. NjBaFAR.

Gamewell Fire-Alarm Telegraph Co. v. Pearce and Jones. Equity Case File 3896. Southern District of New York. NjBaFAR.

Gamewell Fire-Alarm Telegraph Co. v. Mayor and Board of Aldermen of the City of New York. Equity Case File 3370. Southern District of New York. NjBaFAR.

Gold and Stock Telegraph Co. v. Commercial Telegram Co. Equity Case File 2802. Southern District of New York. NjBaFAR.

Gold and Stock Telegraph Co. v. Pearce and Jones. Equity Case File D-1911. NjBaFAR.

Harland, receiver, v. American Rapid Telegraph Co. et al. Equity Case File 3535. Southern District of New York, NjBaFAR.

Harrington, Reiff, and Edison v. Atlantic and Pacific Telegraph Co., and Gould and Others, Executors, &c. Equity Case File 4940. Southern District of New York. NjBaFAR.

LaRue v. Western Electric Manufacturing Co. et al. Equity Case File 3849. Southern District of New York. NjBaFAR.

Page Machine C. v. Dow Jones & Co. Equity Case File 9275. Southern District of New York. NjBaFAR.

Robeson v. American Rapid Telegraph Co. et al. Equity Case File 3352. Southern District of New York. NjBaFAR.

Western Union Telegraph Co. v. Baltimore and Ohio Telegraph Co. Equity Case File 3526. Southern District of New York. NjBaFAR.

Western Union Telegraph Co. et al. v. L. G. Tillotson et al. Equity Case File 2986. Southern District of New York. NjBaFAR.

Western Union Telegraph Co. v. American Union Telegraph Co. Equity Case File 1544. Southern District of New York. NjBaFAR.

Legislative Reports

New York State. Senate. *Report of the Subcommittee of the Committee on Cities Relative to Investigating the Feasibility of Underground Telegraphy in Cities.* 3 May 1882. S. Doc. 82.

———. *Report of the Joint Committee of the Senate and Assembly of the New York Legislature on Telegraphs and Telephones.* Albany: J. B. Lyon Co., 1910.

U.S. Congress. House. *Letter from the Secretary of the Treasury Transmitting a Letter from Professor Morse, Relative to the Magnetic Telegraph.* 28th Cong., 2d sess., 23 December 1844. H. Doc. 24.

———. *Report of the Committee of Commerce on Electro-Magnetic Telegraphs.* 25th Cong., 2d sess., 6 April 1838. H. Rept. 753.

———. *Report of the Committee on Commerce on Electro-Magnetic Telegraphs.* 27th Cong., 3d sess., 30 December 1842. H. Doc. 17.

———. *Report of the Secretary of the Treasury upon the Subject of a System of*

Telegraphs for the United States. 25th Cong., 2d session., 11 December 1837. H. Doc. 15.

U.S. Congress. Senate. *Argument of Patrick B. Delany, Report of Hearings before the Senate Committee on Post-Offices and Post-Roads, May 13 and 20, 1896, Regarding Postal Telegraphy by the Machine System.* 54th Cong., 1st sess., 1896. S. Doc. 291.

————. *Report of the Committee of the Senate upon the Relations between Capital and Labor.* 5 vols. Washington, D.C.: Government Printing Office, 1885.

————. *Report of the Select Committee to Take into Consideration the State and Condition of the Patent Office . . .* 24th Cong., 1st sess. 28 April 1836. S. Doc. 338.

————. *Report of the Senate Committee on Post Offices and Post Roads on Postal Telegraphy.* 49th Cong., 1st sess. 27 May 1884. S. Rept. 577.

————. *Report of the Senate Committee on Railroads on Competing Telegraph Lines.* 45th Cong., 3d sess. 29 February 1879. S. Rept. 805.

Index

BOOKS IN THE SERIES

The Mechanical Engineers in America, 1839–1910: Professional Cultures in Conflict,
by Monte Calvert

American Locomotives: An Engineering History, 1830–1880, by John H. White, Jr.

Elmer Sperry: Inventor and Engineer, by Thomas Parke Hughes (Dexter Prize, 1972)

Philadelphia's Philosopher Mechanics: A History of the Franklin Institute, 1824–1865,
by Bruce Sinclair (Dexter Prize, 1978)

Images and Enterprise: Technology and the American Photographic Industry, 1839–1925,
by Reese V. Jenkins

The Various and Ingenious Machines of Agostino Ramelli, edited by Eugene S. Ferguson,
translated by Martha Teach Gnudi

The American Railroad Passenger Car, New Series, no. 1, by John H. White, Jr.

Neptune's Gift: A History of Common Salt, New Series, no. 2, by Robert P. Multhauf

Electricity before Nationalisation: A Study of the Development of the Electricity Supply Industry in Britain to 1948, New Series, no. 3, by Leslie Hannah

Alexander Holley and the Makers of Steel, New Series, no. 4, by Jeanne McHugh

The Origins of the Turbojet Revolution, New Series, no. 5, by Edward W. Constant II
(Dexter Prize, 1982)

Engineers, Managers, and Politicians: The First Fifteen Years of Nationalised Electricity Supply in Britain, New Series, no. 6, by Leslie Hannah

Stronger Than a Hundred Men: A History of the Vertical Water Wheel, New Series, no. 7,
by Terry S. Reynolds

Authority, Liberty, and Automatic Machinery in Early Modern Europe, New Series, no. 8,
by Otto Mayr

Inventing American Broadcasting, 1899–1922, New Series, no. 9, by Susan J. Douglas

Edison and the Business of Invention, New Series, no. 10, by Andre Millard

What Engineers Know and How They Know It: Analytical Studies from Aeronautical History,
New Series, no. 11, by Walter G. Vincenti

Alexanderson: American Inventor-Engineer, New Series, no. 12, by James E. Brittain

Steinmetz: Engineer and Socialist, New Series, no. 13, by Ronald R. Kline

From Machine Shop to Industrial Laboratory: Telegraphy and the Changing Context of American Invention, 1830–1920, New Series, no. 14, by Paul Israel

Designed by Edward D. King

Composed by Achorn Graphic Services, Inc.
in Berkeley Oldstyle text and display

Printed on 50 lb. Glatfelter Eggshell Offset, A-50
and bound in Holliston Roxite by The Maple Press Company